W9-BMZ-162

About Island Press

Island Press is the only nonprofit organization in the United States whose principal purpose is the publication of books on environmental issues and natural resource management. We provide solutions-oriented information to professionals, public officials, business and community leaders, and concerned citizens who are shaping responses to environmental problems.

In 1994, Island Press celebrated its tenth anniversary as the leading provider of timely and practical books that take a multidisciplinary approach to critical environmental concerns. Our growing list of titles reflects our commitment to bringing the best of an expanding body of literature to the environmental community throughout North America and the world.

Support for Island Press is provided by Apple Computer, Inc., The Bullitt Foundation, The Geraldine R. Dodge Foundation, The Energy Foundation, The Ford Foundation, The W. Alton Jones Foundation, The Lyndhurst Foundation, The John D. and Catherine T. MacArthur Foundation, The Andrew W. Mellon Foundation, The Joyce Mertz-Gilmore Foundation, The National Fish and Wildlife Foundation, The Pew Charitable Trusts, The Pew Global Stewardship Initiative, The Philanthropic Collaborative, Inc., and individual donors.

WALLACE STEGNER

AND THE CONTINENTAL VISION

We are the unfinished product of a long becoming.

—Wallace Stegner, *American Places*

WALLACE STEGNER
AND THE CONTINENTAL VISION

ESSAYS ON LITERATURE, HISTORY, AND LANDSCAPE

EDITED BY CURT MEINE

ISLAND PRESS
Washington, D.C. ● Covelo, California

The editor is grateful for permission to reprint the following previously published material.

From *Conversations with Wallace Stegner on Western History and Literature,* by Wallace Stegner and Richard Etulain. Copyright © 1983 by University of Utah Press. Used by permission of the University of Utah Press. From *Where the Bluebird Sings to the Lemonade Springs,* by Wallace Stegner. Copyright © 1992 by Wallace Stegner. Reprinted by permission of Random House, Inc. From *American Places,* by Wallace Stegner and Page Stegner. Photographs by Eliot Porter. Copyright © 1981 by E. P. Dutton. Edited by John Macrae III. Used by permission of Dutton Signet, a division of Penguin Books USA, Inc. From *Angle of Repose,* by Wallace Stegner. Copyright © 1971 by Wallace Stegner. Used by permission of Doubleday, a division of Bantam Doubleday Dell Publishing Group, Inc. From *Wolf Willow,* by Wallace Stegner. Copyright © 1955, 1957, 1958, 1962, 1990 by Wallace Stegner. Used by permission of Brandt & Brandt Literary Agents, Inc. From *The Sound of Mountain Water: The Changing American West,* by Wallace Stegner. Copyright © 1980 by Wallace Stegner. Reprinted by permission of Brandt & Brandt Literary Agents, Inc.

Library of Congress Cataloging-in-Publication Data

Wallace Stegner and the continental vision: essays on
literature, history, and landscape / edited by Curt Meine.
p. cm.
"The essays contained in this volume were first presented during a three-day symposium held in May 1996 in Madison, Wisconsin"—P. xiii.
Includes bibliographical references (p.) and index.
ISBN 1-55963-537-1 (cloth)
1. Stegner, Wallace Earle, 1909– —Criticism and interpretation.
2. Western stories—History and criticism. 3. West (U.S.)—In literature. I. Meine, Curt.
PS3537.T316Z94 1997
813′.52—dc21 97-14838
CIP

Printed on recycled, acid-free paper

Manufactured in the United States of America

10 9 8 7 6 5 4 3 2 1

To the memory of

HAROLD "BUD" NELSON (1917–1996)

and

GRETCHEN SCHOFF (1932–1994)

CONTENTS

FOREWORD

WALLACE STEGNER wrote in "A Capsule History of Conservation" that "every action is an idea before it is an action, and perhaps a feeling before it is an idea, and every idea rests upon other ideas that have preceded it in time."

Interior Secretary Bruce Babbitt confirmed this observation when, recalling his reading of Stegner's *Beyond the Hundredth Meridian,* he stated that its ideas came to him like a rock through a window, shattering set beliefs. For Babbitt, trained as a geologist, the metaphor was apt. For me, however, Stegner's life and works are more like ingredients in the fine bread my wife makes. She takes them and folds them into the mix, kneading it over and over. Then she covers it, sets it aside, and waits for the bread to rise.

This book is about Wallace Stegner and his own feelings, ideas, and actions. It rises out of a conference held in Madison, Wisconsin, in the spring of 1996, three years after Stegner's death. My wife, our daughter, and I were fortunate to have partaken in that experience. It was a rare gathering. Speaker after speaker, many of whom had personal connections with Wallace Stegner and his longtime partner Mary Page Stegner, confided their thoughts of him, his life, and his work. Often during the three-day conference our eyes moistened as we listened and conversed. Others were with us in spirit: Wendell Berry, Edward Abbey, Aldo Leopold, cherished authors who have written so compellingly about the relationship between people and land. We were in good company.

Apart from this book, the Stegner conference has given rise to other actions. Some are largely personal. Our daughter, for example, has since reflected on her "sense of place" in an article published in our hometown newspaper.

Some are more public. I've taken ideas from the Madison meeting back to Washington, D.C., and folded them into a book my agency, the Natural Resources Conservation Service, has published. *America's Private Lands: A Geography of Hope* is a call to action: to recommit ourselves as a nation to the task of conserving the land. The title comes from Stegner's famous 1960 "Wilderness Letter," in which he encouraged America to preserve its wild places as an important part of our geography of hope. We have taken Wallace Stegner's ideas of wildness and leaned them against our historic tendency to domesticate the landscape. We have revisited Stegner's words. We have sought to define our own "angle of repose" within the landscape. These ideas, gathered up in Madison, have been folded back into our national dialogue about who we are as people and how we are to live with the land.

Whether his ideas are rocks tossed through the window or pinches of ingredients folded into the mix, the words and books of Wallace Stegner have become part of that national dialogue. We are fortunate that he was here. Feelings, ideas, actions: they arise out of others who have preceded them.

To all those who worked to bring these essays and ideas together, thank you. They are a fitting tribute to a good man, and a promising gift to our future.

PAUL W. JOHNSON

ACKNOWLEDGMENTS

THESE ESSAYS were first presented during a three-day symposium held in May 1996 in Madison, Wisconsin. Neither the symposium nor this volume could have come to fruition without the enthusiasm and diligence of the contributors. The editor is especially grateful to them for their quick response to calls and letters in preparing their remarks for publication.

The original symposium was sponsored by the Wisconsin Academy of Sciences, Arts, and Letters and the Academy's Center for the Book. The Wisconsin Academy was established by the Wisconsin State Legislature on March 16, 1870, to encourage investigation in the sciences, arts, and letters and to disseminate information and share knowledge. The Stegner symposium was very much in keeping with the interdisciplinary vision of the academy's founders. The Academy's Center for the Book, which is affiliated with the Center for the Book at the Library of Congress, was established in 1986 to promote appreciation of books, book arts, reading, and the written word.

Symposium endorsements came from the Aldo Leopold Foundation; the International Crane Foundation; the State Historical Society of Wisconsin; the University of Wisconsin–Extension; the University of Wisconsin–Madison Departments of English, Geography, History, Institute for Environmental Studies, and School of Journalism and Mass Communication; The Wilderness Society; the Wisconsin Arts Board; and the Wisconsin Humanities Council.

Financial support from the following organizations and people made the symposium possible: the Aldo Leopold Foundation, Karl Andersen and Carolyn Heidemann, A. Roy and Pat Anderson, Marvin and Ellouise

Beatty, Allan and Margaret Bogue, Stephen Born, Reid and Frances Bryson, Martha Casey, David and Jean Cronon, John Dahlberg, Emily Earley, Jonathan Ela, the Evjue Foundation, William and Betty Fey, Richard and Susan Goldsmith, Inga Brynildson Hagge, John and Patricia Healy, Michael Hinden, Willard and Frances Hurst, Henry and Annrita Lardy, Ronald and Margaret Mattox, John and Barbara Mueller, Eleanor B. Mulvihill, Harold and Ann Nelson, G. B. Rodman, Jenifer and Edmund A. Stanley Jr., Nancy Townsend, Gerald and Marion Viste, and Thompson and Dianna Webb.

Many contributions to the symposium (including several anonymous donations) were made in memory of Harold "Bud" Nelson and Gretchen Schoff. Harold "Bud" Nelson contributed importantly to the work of the symposium planning committee. An internationally respected professor of journalism at the University of Wisconsin–Madison, Dr. Nelson was also a dedicated conservationist, with interests ranging from prairie restoration to wilderness protection. After enjoying a correspondence with Wallace Stegner that began in 1984, Nelson initiated the effort to bring Stegner back to the University of Wisconsin–Madison for an honorary doctorate in 1986.

Gretchen Schoff, an early member of the symposium planning committee, was professor of environmental studies, general engineering, and integrated liberal studies at the University of Wisconsin–Madison. Dr. Schoff's career was distinguished by the ease with which she crossed disciplinary boundaries. She was particularly interested in the humanistic aspects of environmental issues, a concern reflected in her deep appreciation of the work of Wallace Stegner. She interviewed Stegner during his 1986 visit to Madison, and her memoir of that interview, "Where the Bluebird Sings: Remembering Wallace Stegner," was published in the collection *The Geography of Hope: A Tribute to Wallace Stegner* (San Francisco: Sierra Club, 1996), edited by Page Stegner and Mary Page Stegner.

This volume is dedicated to the memory of these two special friends and Stegner enthusiasts.

The symposium could not have taken place without the backing of officers and staff of the Wisconsin Academy of Sciences, Arts, and Letters, in particular senior associate director Richard J. Daniels, special events coordinator Gail Kohl, and graphics designer Marty Lindsey. Academy editorial director Faith B. Miracle served as conference facilitator and indispensable liaison between the academy and the organizing committee. Tom

Uttech gave permission for use of his painting *Moonrise, Baptism Lake* (and details from it) in developing the visual theme of the symposium.

The symposium planning committee consisted of Pat Anderson, Allan Bogue, Nancy Langston, Charles Luthin, Curt Meine, Faith B. Miracle, Harold "Bud" Nelson, Walter Rideout, and Gretchen Schoff. Conference coordinator Ann Ostrom guided participants smoothly through the event. Special thanks to Wisconsin Academy president Ody J. Fish, University of Wisconsin president Katherine Lyall, Academy Center for the Book president Richard Telfer, Professor Nancy Langston, Professor Emeritus Walter Rideout, and discussion leader Catherine Rasmussen for their contributions to the symposium program. Fred Ott provided books from his collection of Stegner first editions for exhibition during the symposium. Charles and Nina Leopold Bradley were staunch supporters of the idea from the outset and graciously extended their hospitality to all during the gathering.

Barbara Dean of Island Press expressed strong interest in this project from its inception and responded quickly to many inquiries in the course of its completion.

Finally, a thank-you to Mary Stegner and Page Stegner for their endorsement, encouragement, and good wishes.

INTRODUCTION

When word went out in the spring of 1993 that Wallace Stegner had died from injuries suffered in an automobile accident in Santa Fe, New Mexico, his many readers were caught off guard. Although Stegner was eighty-four at the time of his death, interest in his work was again on the upswing. His final years had seen publication of *Crossing to Safety* (1987), *The American West as Living Space* (1987), *The Collected Stories of Wallace Stegner* (1990), and *Where the Bluebird Sings to the Lemonade Springs* (1992), books that confirmed the loyalty of his veteran readers while gaining him legions of new admirers. For more than half a century Stegner's work had stood out, monumentally, on the literary landscape. When he died, the effect was something akin to the explosion of a Mount St. Helens: for many a reader, a familiar literary landmark was simply gone.

In the aftermath, many who knew Stegner—through his words, his works, or his personal influence—paused to mark the event, survey the field, and look for signs of continuity. Since 1993, Stegner has been the focus of several symposia, a biography, a documentary film, and several volumes of essays.[1] The essays collected here were first presented in Madison, Wisconsin, in May 1996 at a symposium held under the auspices of the Wisconsin Academy of Science, Arts, and Letters. The symposium's goal was to explore the significance of Stegner's life—its inspirations and limitations, its challenges and discouragements, its insights and continuing relevance. This collection presents those explorations to a broader audience.

Certainly, Madison was an appropriate vantage point from which to review Stegner's legacy. Stegner taught at the University of Wisconsin in the late 1930s, years that brought him important friendships, new personal and professional challenges, and a wider context in which to explore his characteristic themes. In his conversations with Richard Etulain, Stegner noted that in Madison he "was made conscious of the outside world in ways that I had never been before. . . . Out of it came these questions of what one owes the society, and where does the individual fit in? . . . A lot of ideas came into my head that had simply never lodged there before."[2] The immediate impression of these years can be found in Stegner's early novel *Fire and Ice* (1941).[3]

The mark of those Wisconsin years remained strong. In Stegner's last novel, *Crossing to Safety*, he revisited his Madison experience in tracing the course of close personal friendships that began there. In May 1986, even as Stegner was completing the manuscript of *Crossing to Safety*, he returned to Madison to receive an honorary degree from the University of Wisconsin. During this same period he composed his essay "The Sense of Place" under the sponsorship of the Wisconsin Humanities Committee. (The essay was later reprinted in *Where the Bluebird Sings to the Lemonade Springs*.)[4]

Madison was an appropriate venue for cultural and geographical reasons as well. Stegner's lifework drew upon rich traditions in American history and the American conservation movement that are important parts of Wisconsin's heritage. Over the decades, the University of Wisconsin has nurtured the likes of John Muir, Frederick Jackson Turner, and Aldo Leopold, all of whom ranked as key influences in Stegner's own writing and conservation career. Less directly, but nonetheless fittingly, the midwestern setting provided an opportunity to consider Stegner's impact beyond the region—the American West—with which he is most closely identified. Through his history and fiction, Stegner explored the connections and tensions between the arid, expansive, hopeful, misunderstood, and problematic American West and the wetter, wealthier, more settled, more Europe-oriented, and culturally dominant East. His personal geography reflected this theme, as he and his family migrated yearly between the coastal range of central California and the green hills of Vermont. During these migrations, Stegner might have paused and thought of Wisconsin's position in the midcontinent as a place to look both ways from the middle. This geographic orientation gave the symposium its theme and gives this collection its focus.

▽ ▽ ▽

STEGNER'S LIFE was characterized by his constant ability—*need* may be more apt—to achieve in his art the largeness he first apprehended as a boy on the short-grass plains of Saskatchewan. "Largeness is a lifelong matter," he wrote in 1988. "You grow because you are not content not to. You are like a beaver that chews constantly because if it doesn't, its teeth grow long and lock. You grow because you are a grower. You're large because you can't stand to be small."[5] The inner drive to gain in experience, knowledge, and perspective (and, just maybe, if one blends these well, wisdom) can be easily skewed, but it cannot be easily directed or dissuaded—a dilemma Stegner scrutinized in much of his fiction and nonfiction.

In *The Uneasy Chair* (1974), his masterful biography of his friend and mentor Bernard De Voto, Stegner converts this dilemma into an organizing attribute. De Voto's path in life paralleled in many ways Stegner's own, especially in its sensitivity to the East–West dynamic in American history and life. As a result, Stegner was able to observe with skill and sympathy De Voto's eventual grasp of "the grand theme of the development of the continental nation."[6] That comprehension reflected De Voto's own innate need to grow socially, intellectually, and artistically. Stegner wrote of a period of personal crisis early in De Voto's life when, despite emotional upheaval and discouragement, De Voto was nonetheless able to gain for the first time a sense of the continuities, in time and space, that bind the continent. "His continental perspective," Stegner wrote, "had expanded into a vision."[7]

Wallace Stegner likewise strove to achieve, in his writing and in his life, a continental vision. By the 1970s, critics had begun to describe him as the "dean of American western writers." He gained that reputation through his thorough immersion into the history, literature, and developing culture of the region he knew best. His untiring efforts to shape and reshape our understanding of the West may very well stand as one of twentieth-century American literature's great accomplishments. Yet Stegner's West is not an isolated region detached from the rest of the continent or the world. The forces and trends he described, recorded, dramatized, and criticized in the West did not arise solely from within but were bound to forces and trends elsewhere. He saw that the western landscape allowed powerful and pervasive forces in American society—the striving for political independence and political influence, the relationship between collective action and individual initiative, the power of knowledge and the power of wealth, the appreciation of wildness and antipathy toward it—to reach points of ultimate expression. The West, Stegner wrote in his essay "Making a Myth," is "not a marginal region but the mainstream, America only more so."[8] In a lifetime of working through the meaning in that sentence, Stegner became a great writer not only of the West but of the continent.

One need not read much into Stegner before sensing that he understood his own expanding vision less as an accomplishment than as a compulsion and a responsibility. He suggested time and again that to grow as a writer, to find one's personal angle of repose, and to be a decent citizen in an evolving democracy, one must try to understand the relationships connecting history, culture, and landscape. In this way we might understand ourselves better, place ourselves, and tell our stories more fully and faithfully—and, in our civic lives, begin to take the necessary steps toward permanence as dwellers within a landscape.

Stegner's continental vision was not without its blind spots. He himself acknowledged, for example, his failure to document in greater detail the experience of Native Americans; his too quick willingness to accept certain federal initiatives early in his conservation career; his sometimes gruff response to the countercultural tumult of the 1960s. It is equally evident, however, that Stegner worked throughout his life to write and think of his culture in the broadest possible terms. Moreover, what he did not write about personally he could appreciate in such students as Edward Abbey, Wendell Berry, Ernest Gaines, and M. Scott Momaday. And Stegner, as much as any of his contemporaries, succeeded in bringing the realms of literature and the natural world back into some state of conversation. The vision remains, as Stegner might have put it, not a fact but a task. In Dorothy Bradley's words from this volume, Stegner's work obligates us "to clarify the task, translate the past, define the next meridian, and articulate the hope."

Taking the measure of Stegner calls for varied perspectives. The contributors in this volume represent a wide spectrum of disciplines and fields: literature, history, law, geography, conservation biology, women's studies, politics. They comprise academics and nonacademics; those who knew Stegner well and those who knew him only on the printed page; acknowledged Stegner scholars and newer voices. They offer a fair representation of the continent's landscapes as well: the driftless hills of northeast Iowa, Colorado's Front Range, California's southern coast, Oregon's coast ranges, the New Hampshire seaboard, the Snake River below Hell's Canyon, the outliers of Arkansas' Ozarks, the dunes of Lake Michigan's southern shore, Montana's Bitterroot Valley, the moraines of southern Wisconsin, the lower reaches of the Potomac, the upper reaches of the Missouri, and

Utah's Salt Lake. If Stegner was a regional writer, he was one of cosmopolitan appeal.

▽ ▽ ▽

FROM TIME TO TIME STEGNER referred to the need for a "usable" and "possessed" past—a body of stories and histories that could allow westerners and Americans in general to orient themselves in the present and shape a sound future. At a time when many are coming to appreciate the need for an understood past and place, Stegner continues to offer readers an unusual blend of strength, encouragement, realism, and sobriety. His effectiveness as a writer came as much from what is in the lines—his commanding prose style—as from what is between them—his evident respect for his readers.

Stegner understood that we are always in the process of inhabiting our continent, and in a sense our lives. By defining for us his vision of this time and place as part of a continuing process, "an unfinished product," he opened us to the possibility of renewal. And he did so with wit, vulnerability, and humanity. Stegner has left us a large legacy. The essays in this volume seek to embrace that legacy and respond to it with the same respect that Stegner showed his readers.

PRELUDE

STEGNER IN PERSPECTIVE

THE INTENT OF THIS VOLUME is not simply to celebrate Stegner but to explore his work and its impact. The contributors have been encouraged to explore those facets of Stegner's work they personally found most illuminating and provocative. The essays have been arranged within three broad categories: literature, history, and conservation. These of course are not altogether discrete categories in Stegner's case. In the Prelude to the book, therefore, Charles Wilkinson orients us amid these overlapping fields by noting the qualities in Stegner's writing that have made him so influential a force in our perceptions of the American West: his authenticity and originality; his fundamental understanding of biogeographic conditions and social dynamics; his fusion of varied realms of knowledge to comprehend regional complexity; his blend of scholarship, storytelling, and conservation action. While especially germane to Stegner's work on the West, these qualities are truly of continental relevance.

CHAPTER ONE

WALLACE STEGNER AND THE SHAPING OF THE MODERN WEST

CHARLES WILKINSON

In the middle 1960s, in the way that life's passions can arrive unannounced and full-blown, I began a love affair with the American West. I was practicing law in Phoenix and the matchmaker was Paul Roca, an expansive Arizona lawyer. Paul told me stories of Fray Marcos de Niza, the Mogollon Rim, ironwood, and crooked land deals.

I became a bookseller's dream. For more than a decade I read widely and feverishly—and erratically and inefficiently—about the West. I had learned specifics, but had no context. Then, in 1977, I first read Wallace Stegner. *Angle of Repose* was the book. I was entranced by his plain words and the scope and depth of what he knew. I bought an armload of Stegner books and dove in.

I was teaching at Oregon at the time. Public land law, Indian law, and water law—western courses all—were my subjects. Yet Stegner's writing was telling me more about the law than the court decisions and the statutes. Law, I began to realize, grows out of a society. To learn about law, learn about the society. Law, in other words, has a habitat. And to learn what a society's law ought to be, measure it against the society's history, peoples, lands and waters, possibilities, and limitations. Reluctantly, I decided not to cancel my reading of pallid law opinions, interminable treatises, and sterile codes. But I did come to understand that I should give at least equal time to ranches, irrigation ditches, mine shafts, crisp air, ridgelines, canyons, chuckwallas (which I privately suspect may have been Stegner's favorite animal), and the crawl of subdivisions out into the open plains.

In 1984, I sent Stegner some of my writing and asked if we could get together. The request struck me as fruitless and perhaps foolish: he'd never answer. To my astonishment, I received an enthusiastic response by return mail. I didn't know then that he'd answer anybody's letter, usually promptly, always with a generous spirit. It is fortunate for posterity that hardly anyone learned the secret I'd come upon by happenstance. Mary and Wally would have had to pave over part of their sublime meadow in the Los Altos foothills, and *Crossing to Safety* might never have seen the light of day.

My letter, born of such modest expectations, led to one of the largest blessings of my life. Wally and I became good friends and I loved him, not just for his writing, but also for the way he was as a person, for his clarity, fire, and constructive stubbornness, and for his personal grace and dignity. Taking me under his wing as he did was all the sweeter, ultimately, because he did it for so many.

By 1994, I'd long wanted to teach a full course on Stegner. His writing, I thought, would be of interest to anyone. He wrote in luminous prose of the most basic needs, pains, pleasures, and aspirations of the individual, family, and society. He told deeply, longingly, of place—any place, so long as you became sufficiently embedded in it to make it *your* place.

While Stegner wrote of the human condition, broadly conceived, most of his writing was set in the American West because that was the region he knew. He wrote, I believe, more serious words, and certainly more valuable words, about the West than anyone ever has. He had become an opinion leader, one of the most influential people in the twentieth-century American West. And immersing students in Stegner was the perfect way to teach them about the West. Before I met with my dean to propose the Stegner seminar, I had rehearsed my pitch about how all of this might not seem like law, but actually was. I never needed it. "Great," he said. "Let's do it." Our conversation turned to other topics, and then we finished up. I was almost out the door when he said, "Oh, Charles, by the way, I read that copy of *Beyond the Hundredth Meridian* that you just happened to give me a couple of weeks ago."

So in the winter of 1995 I began the Stegner seminar with an enthusiasm that exceeded any course I've ever taught. The seminar was cross-listed in both Law and English, and the students were diverse and motivated. They read *The Big Rock Candy Mountain* over Christmas break and then *Conversations with Wallace Stegner,* Richard Etulain's book of long interviews. For the third session, we did *Wolf Willow* and, notably, "Molly's Story," which caused me to understand some new things about Stegner and his hold on people.

The week before the *Wolf Willow* class, one of my students asked if his mother could sit in. She was a rancher and keenly interested in Stegner. I met her, tall, gray-haired, and fit, before class.

The students talked about *Wolf Willow.* Several commented on the force of Stegner's poignant return to Whitemud—the fictitious name for his boyhood Saskatchewan town of Eastend—and the smell of wolf willow, the "pungent and pervasive . . . smell that has always meant my childhood. . . . It is wolf willow, and not the town or anyone in it, that brings me home."[1] In discussing another part of the book, a Lakota student, expert in American Indian history, explained that although he had never known much of the Métis in Canada, their story was the story of the Plains Indians in the United States as well.

We turned to "Genesis," one of Stegner's own favorite stories. He had wanted to read it for audiotape, but he didn't get to it and so his son, Page, did it for him after Stegner's death in 1993. "Genesis" is the long and dramatic story of Saskatchewan's killing winter of 1906–1907 and of Ray, the foreman; the veteran ranch hands, Spurlock and Panguingue; and Rusty Cullen, a young English boy who rescues the foul-tempered Spurlock from sure death in the blizzard.

We talked in class about whether the kind of cooperation that was the hallmark of "Genesis" was a real thing on western ranches. Fairly late in the class I asked, "Mrs. Rogers, what do you think? Is 'Genesis' overdrawn? Could it really have happened that way?"

I realized that Mrs. Rogers—though you wouldn't have known it from the way she sat so quietly during the class—was aching to pitch in. She explained that her father took up ranching in Montana in the early 1890s when memories of the epic winter of 1886–1887 were still fresh. As a young girl, Mrs. Rogers had heard many stories of that winter, much like the one that hit Saskatchewan two decades later.

Stegner, she said, was "right on the mark. That was exactly how it was." She then went on to dissect the web of duty and cooperation that she thought "Genesis" stood for. Mrs. Rogers told us how ranch country has changed, how many working ranches have gone to hobby ranches or, worse yet, subdivisions. But there are still many places, she concluded, where the spirit in "Genesis" lives.

Mrs. Rogers wanted to revisit the observations a couple of students had made about Molly Henry, the wife of Ray, the foreman. In "Carrion Spring," Stegner's extension of "Genesis," Ray talked Molly into staying with him, even after that horror of a winter, in that Canadian cattle country instead of going back to town in Malta, Montana. Ray rescued a coyote pup, one of the winter's few survivors, and gave it to her. Some students wondered if Molly had shown weakness in staying in that country with her husband.

Mrs. Rogers said that Molly was authentic, but not weak. Molly loved her husband and wanted to raise a family with him. Molly, Mrs. Rogers believed, had strength aplenty and did backbreaking work around that ranch.

We had gotten deep into Stegner, ranch country, cooperation, and western women. It was about the best class I'd ever taught—or, more accurately, convened. At least it was the best until the next one.

A week later, I picked up my mail in the morning. The seminar met at 4:00 in the afternoon. I found a letter from Mrs. Rogers. I opened it to find this note:

> Dear Professor Wilkinson,
>
> Thank you very much for allowing me to audit class last Tuesday. As I'm sure you could tell, I thoroughly enjoyed the opportunity, and reflected upon it as I returned to Parker. During the drive I found Molly was with me, appreciative of the comments I'd made about her. Before she evanesced, she confided in me something of what her blizzard experience had been. I have tried to capture her comments and I forward my transcript to you, thinking they might be of interest.
>
> Sincerely yours,
> Rowena Rogers

Attached to the note was "Molly's Story." It was a tough, sweet, evocative remembrance of Molly's time, alone, at the ranch house during that dreadful winter. Wood was in short supply, so she moved her mattress into the kitchen, which she kept heated. She had to chop ice from the frozen creek for water. The hog died, and so did the calf. She kept the cows alive, and most of the chickens, which she ate, along with the calf liver—but not the hog liver. There were many nights when Molly was sure that Ray, the handsomest man she ever knew, was dead.

Molly told and retold the whole story of that winter to her children. "Molly's Story" is too long to present here, but it included this:

> "One night," her telling of the story would continue, "when I was boiling up the next-to-last chicken, I heard an awful noise on the porch. I thought all the snow must have slid off the roof and was wondering if ever I'd be able to open the door again when it opened itself. An awful looking monster came in. I grabbed the poker to hit it with, but it grabbed me before I could." This was the part of the story the children came to love best.
>
> "Tell us again, about the monster, Mommy," they'd scream with delight.
>
> "Well," she'd tell them, "it was a lot bigger than me, and it had this terrible brown hairy face, with little squinty red eyes, and awful bloody, cracked lips that only half covered its big white teeth. I thought it was going to bite me, so I swung at it with the poker, but it just pulled me toward it—and . . ."
>
> "What, Maw, what, Maw?" This was after the story had become a ritual.
>
> "It kissed me."
>
> "It was Daddy!" Reassured once more that he'd made it safely home, they'd jump up and down and race around the room. "And then what?"
>
> "And then," she had become a skillful storyteller, "we moved back to Malta where it was safe and we'd never have to lose each other in another blizzard."
>
> "No, no, you didn't. He gave you a coyote pup and you said you'd stay."
>
> She thought of that pup, the only life among all that death around them, except for the new life she knew was inside her. She had to keep the pup alive. Even now she still had a little hidden hurt from when it'd learned to kill chickens and she'd shot it without asking Ray to do it for her. He helped her when he found her trying to dig a grave.
>
> "Wild ones," he said. "It's not a kindness to save them."
>
> A kindness to me, she thought, but by then she had Ray Junior to love.

> "That's right," she told the children. "We saved the coyote pup, and its children are still out there singing you children to sleep, and its time for me to blow out the lamp."

Near the end of class that day, I handed out copies of "Molly's Story" to the students, who read it, rapt, a sacred gift from Mrs. Rogers, and Stegner, too.

▽ ▽ ▽

IF STEGNER'S DEPICTIONS of the West were utterly authentic, it is because he saw so much of it. Saskatchewan in the 1900s was frontier—about like eastern Oregon, Idaho, Wyoming, or eastern Colorado in the 1870s or 1880s. Stegner then saw the West of the 1910s and 1920s, involuntarily, dragged around by his father. The family's stopping points included both rural towns and the cities, such as they were, of Great Falls, Seattle, Reno, and Salt Lake City. He lived in or near the Great Plains, the northern Rockies, the Cascades, the Sierra Nevada, the Great Basin, and the Colorado Plateau. Stegner was among the few to see Glen Canyon before the inundation. During most of his adult life, he lived in the Bay Area and witnessed the rise of the modern industrial West.

His eyes saw the whole thing: from early sodbuster days to what he called Denver's "ringworm suburbs."[2] He worked his old typewriter, pecking out the truth his eyes had seen, and it was truth, even if those pages gleamed with literary genius.

And so you begin with the integrity of authenticity.

▽ ▽ ▽

STEGNER'S BODY OF WORK was not only authentic but original. It regularly left us with a sense of discovery. *Beyond the Hundredth Meridian* did this in a number of respects. John Wesley Powell, evicted from public life amid the boos of the Irrigation Congress in Los Angeles in 1893, had faded from memory—from history, really. He was remembered vaguely as a daring canyon explorer. Yet Powell, who was a leading national figure in his own time, understood the West as well as anyone of his century. The ideas in his 1878 *Arid Lands Report* challenged the premises about the region and deserve to be debated yet today. Stegner revived the memory of Powell.

He resurrected, too, the mostly forgotten events that Powell instigated. In 1889, Major Powell, who at the time headed both the U.S. Geological Survey and the Bureau of Ethnology, talked the General Land Office into withdrawing essentially all western public lands from settlement. Existing operations could continue, but Powell had shut down the public domain from the 100th meridian to the Pacific. The lands would remain in deep freeze until Powell finished his Irrigation Survey and made his recommendations about what to do with them. In the meantime, Manifest Destiny would give way to restraint, patience, study, and planning. Powell's project was unprecedented, not to mention un-American and, worse yet, unwestern.

Of course, Powell's Grand Plan had no chance, at least viewed in retrospect. But it was the law for more than a year, during which time it was the focal point of national lands policy, before western senators voided the withdrawal by statute. Other than Theodore Roosevelt's set-asides of nearly 150 million acres of land for national forests in the early twentieth century, no single episode better epitomizes the essential conflicts in public land policy than the Powell withdrawal and its aftermath, events once nearly forgotten but restored to the national memory by Wallace Stegner:

> But they hadn't given him time. They had beaten him when he was within a year of introducing an utterly revolutionary—or evolutionary—set of institutions into the arid West, and when he was within a few months of saving that West from another half century of exploitation and waste. It was the West itself that beat him, the Big Bill Stewarts and Gideon Moodys, the land and cattle and water barons, the plain homesteaders, the locally patriotic, the ambitious, the venal, the acquisitive, the myth-bound West which insisted on running into the future like a streetcar on a gravel road.[3]

Bruce Babbitt, interior secretary in the Clinton administration, explained his own experience with *Beyond the Hundredth Meridian*. Growing up in Flagstaff, Arizona, a traditionally western timber and grazing town, he had, he realized later, a "closed in, confined childhood." Then one day, browsing in a bookstore, he found the Powell book. It was, Babbitt said, "the rock that came through the window," opening a bright, new, outside world of ideas.

Stegner's rediscovery of the life of Mary Hallock Foote—his Susan Burling Ward in *Angle of Repose*—is another example of the sense of exploration that readers find in Stegner's books. Among its many other virtues, the book is fresh and vital with the energy of a different West, one seldom told, a place for a talented and independent woman.

Stegner's spirit of discovery plays out in at least two ways. One is his own unearthing of lost historical treasure. The other is his art of passing on to others the context for their own discoveries—whether a future secretary of the interior with his own spirit of inventiveness or a ranchwoman willing and able to reach for a story within a story about an earlier ranchwoman named Molly.

▽ ▽ ▽

STEGNER HELPED DEFINE the West. He articulated, as no one ever had, the exact point of passage across the dry line into the American West. Just a few words from Bruce Mason's long drive west in *The Big Rock Candy Mountain* present this defining moment:

> At the next service station where he stopped he felt it even stronger, the feeling of belonging, of being in a well-worn and familiar groove. He felt it in the alacrity with which the attendant shined up his windshield and wiped off his headlights and even took a dab at the license plates, in the way he moved and looked, in the quality of his voice and grin. Anything beyond the Missouri was close to home, at least. He was a westerner, whatever that was. The moment he crossed the Big Sioux and got into the brown country where the raw earth showed, the minute the grass got sparser and the air dryer and the service stations less grandiose and the towns rattier, the moment he saw his first lonesome shack on the baking flats with a tipsy windmill creaking away at the reluctant underground water, he knew approximately where he belonged.[4]

Later, Stegner would place the edge of the West at the 100th meridian. Of course, this kind of thinking had been around a good while. Walter Prescott Webb expounded on the dry line at length, using the 98th meridian as the marker, once even calling it "that magic line."[5] But it was Stegner who captured the public's understanding and imagination, using, as he did, the 100th meridian in the title of the West's greatest book.

The 100th meridian is now a byword, and it turns out to be good science as well as good symbolism. Biologist Carl Bock explains that the 100th meridian is a dividing line for many kinds of birds. Not only does it define species habitat for "eastern" birds and "western" birds, but it often makes even finer distinctions, such as varieties within species. Bock's studies show that red-shafted flickers live west of the line, while yellow-shafted flickers live to the east. In all, Bock believes that the 100th meridian is "the most important biogeographical boundary in North America." Another biologist at the University of Colorado, David Armstrong, says that his studies of mammals show the same results as Bock's research on birds. Armstrong, who, like Bock, emphasizes Stegner's role in focusing attention on the 100th meridian, concludes that it marks the beginning of the "ecological west."[6]

The West's aridity, like the 100th meridian that roughly defines it, has long been acknowledged. But again it was Stegner who explored it most deeply, and from the most angles, and who popularized it—not through flash or slogans, but through authenticity. He explored aridity through the eyes of Bruce Mason, John Wesley Powell, William Gilpin, Susan Ward, Oliver Ward, Brigham Young, and many others, including, in his essays, his own. In doing so, he added still more definitional substance to the West and gave the region a fuller sense of its past and its possibilities.

His portraits of the West included the region's peoples. He had no truck with, as he variously called them, the boomers and boosters, the "snarling states-rightists," the plunderers, and the Gilpins who made up the "mythbound West." But he wrote with compassion, respect, and often love of the subcultures who had not been given much due in the serious history and literature. *The Gathering of Zion* is a grand, definitive history of the Mormon Trail. *Mormon Country,* more than any other book, explains in a fairminded way the tangle of community, isolation, family, prejudice, warmth, zealotry, hardheadedness, and devotion that makes up the Church of Latter-day Saints. Stegner, who admired the Wobblies as early as the 1920s and believed in their cause, wrote *Joe Hill,* a historical novel about one of their leaders. And Stegner, whose heroic mother made him feel worthy through all his father's storms, long ago was what we now call a feminist. Strong women pervade his work. Implicitly he shows that the West was settled not by lone mountain men or miners heading out toward the horizon, but by families.

One Nation, published in 1945, deals explicitly with the dispossessed and was far ahead of its time in its treatment of race. I wish Stegner had written more about Indians, but his exploration of the Métis in *Wolf Willow*—respected by Canadian scholars as well as my Lakota student—dug deeply into their culture. Their intelligence is manifested in their land system that gave stream frontage to all lots. This system, Stegner observed, was "far better adapted to the arid and semi-arid Plains than the rectangular surveys were."[7] He also admired their esprit, as seen in their colorful songs, dances, and clothing.

Stegner's definition of the West included the land, always the land. In every book the land is a character, many characters really. On any given page of Stegner's writing you may meet sandstone, limestone, granite, or uplift; elm, pine, oak, mesquite, or cottonwood; bison, deer, weasel, spider, grizzly, or merganser; whirlpools, rapids, cloudbursts, or shafts of light; or a V-shaped mountain pass. Or long, open space, especially space:

> In the West it is impossible to be unconscious of or indifferent to space. At every city's edge it confronts us as federal lands kept open by aridity and the custodial bureaus; out in the boondocks it engulfs us. And it does contribute to individualism, if only because in that much emptiness people have the dignity of rareness and must do much of what they do without help, and because self-reliance becomes a social imperative, part of a code.[8]

The land in Stegner's West is sometimes inspirational, often mundane, always organic with human life. On one level, Mexican Hat, in southern Utah, is just barren terrain: "The space that surrounds Mexican Hat is filled only with what the natives describe as 'a lot of rocks, a lot of sand, more rocks, more sand, and wind enough to blow it away.' "[9] Yet Mexican Hat is also on the Colorado Plateau, a place of long views that "fill up the eye and overflow the soul."[10]

Stegner's view of the West created a new body of knowledge because his way of looking at the West was so multifaceted. It brought so much information to bear, information that had never before been welded together. He saw the whole: history, geography, culture, economics, law, biology, beauty, human emotion, and doubtless more, all at once. Page Stegner puts it best:

> But more important, I think, than any formal instruction he offered was his habit of imbuing virtually everything he came across with substance. My father, for example, could never just *look* at scenery. If we happened to be driving across the Colorado Plateau through southern Utah, say from Cisco to Price along the Book Cliffs, he'd offer up an anecdote about Powell being rescued by Bradley in Desolation Canyon, and then explain to his slightly annoyed eight-year-old boy (me), who was trying to concentrate on his Batman comic, who Powell was and why he was important. Then he'd point out the La Sals

> and Abajos to the south and tell that boy something about laccolithic domes, betting him he couldn't spell laccolithic. He'd comment on the immensity of geological time and the number of Permian seas responsible for the deposition of the Moenkopi, Chinle, Wingate, and Kayenta formations (he could identify them all) on our left and the Dakota sandstone and Mancos shale on our right. He'd observe the Fish Lake Plateau far to the west and remember something of his boyhood summers at that lake, though he was never particularly loquacious about his own childhood except in his writing. Crossing over the Wasatch Plateau and heading south through the Spanish Fork canyon would remind him of the specific dates of the Escalante/Dominguez expedition through the region (September 23, 1776), and that it was exactly fifty years before Jedediah Smith came through following essentially the same route. He had a kind of holistic relationship with the land, and he couldn't look at it without remembering its geological history, its exploration, its social development, its contemporary problems, and its prognosis for the future.[11]

Stegner's West, then, is an enormously complicated place. He found that transcendent truths emerged from the data he painstakingly collected from what he saw and read. When evoked through his prose, at once straightforward and elegant, the truths emerged incontestable and vivid.

The truths emerged from his fiction every bit as much as from his histories, biographies, and essays. *Angle of Repose* tells hard truths about aridity and the limits of the land—and of human ingenuity—in Oliver and Susan Ward's failed endeavors in the heat and dust of the Boise Valley. Rootlessness can be an abstraction until you read *The Big Rock Candy Mountain* and meet a boy whose father had torn him out of his place; the boy had to sell his invalid little colt with the braces on its legs, supposedly to a man who would care for him. On the way out of Whitemud, the family drove past the dump:

> So they left town, and as they wound up the dugway to the bench none of them had the heart to look back on the town they were leaving, on the flat river bottom green with spring, its village snuggled in the loops of river. Their minds were all in the bloated, skinned body of the colt, the sorrel hair left below the knees, the iron braces still on the broken front legs.
>
> Wherever you go, Elsa was thinking, whenever you move and go away, you leave a death behind.[12]

People were always on the move, and the land was always being assaulted, Stegner believed, because of the mythmakers who cried out that there was a better place over the ridge and that the land could be conquered there. This of myth:

> Verifiable knowledge makes its way slowly, and only under cultivation, but fable has burrs and feet and claws and wings and an indestructible sheath like weed-seed, and can be carried almost anywhere and take root without benefit of soil or water.[13]

To combat the mythmakers, the West had to move beyond individualism to cooperation and then to true society:

> Civilization is built on a tripod of geography, history, and law, and it is made up largely of limitations.[14]

Stegner thought that the natural world should be protected for its own sake, but he wrote and talked more about another dimension. The way we treat the land is a measuring stick for the quality of our society, for the level of morality in our civilization. In "Wilderness Letter," his most noted piece on the environment, he raged against the loss of the natural world. But a main part of the loss was to us human beings, and in a larger sense to the civility that makes a society:

> Something will have gone out of us as a people if we ever let the remaining wilderness be destroyed. . . . An American, insofar as he is new and different at all, is a civilized man who has renewed himself in the wild. . . . We simply need that wild country available to us, even if we never do more than drive to its edge and look in. For it can be a means of reassuring ourselves of our sanity as creatures, a part of the geography of hope.[15]

This body of work is composed of a broad, thick base of data and an elaborate philosophy built upon it. Although Stegner is quotable—eminently quotable—that is not the essence of his literature. Much of his writing is plain and detailed: reports of work and family, day-to-day lives. When asked "which Stegner book should I read?" I can offer up any number of great books, but I can't single out any of them that seriously begins to set out Stegner's view of the West. Stegner's West is far too full for that.

Even his most-quoted passage, composed of so few words, in fact opens up a long vista, canyon after canyon, ridge after ridge, to the far horizon.

Yes, the West is the "native home of hope." But the irony is that the hope of the West has most often been the quest of Bo Mason, and millions of others, for the big rock candy mountain and the place where the bluebird sings to the lemonade springs. Yes, "cooperation, not rugged individualism," most "characterizes and preserves" the West. But Powell, the cooperator, went down to the knives of Gideon Moody and Big Bill Stewart, and the rugged individualists of Utah still stonewall wilderness in Stegner's beloved canyon country. Yes, maybe we can "create a society to match its scenery."[16] But if you understand Stegner's thinking, you know that he thought the possibilities might be scant, and waning. You have to walk his literary canyons and ridgelines to understand the full texture of those words. Like all great hikes, it is well worth it, but a long hike nonetheless.

I SAID AT THE START that Stegner has become one of the most influential people in the twentieth-century West. Of course, that is something I can't pretend to prove in some final sense. It is impossible to measure the changing of minds, the prelude to the changing of a society.

Yet it seems to me a fair observation. Even today, the figures who most clearly molded the modern West came early in the century—Muir, Roosevelt, Pinchot. But who else would rival Stegner's influence? The most productive legislators—Mansfield, Church, Hatfield, Aspinall, Jackson of Washington? The activists—Chávez, Brower? The leading interior secretaries, holders of the region's most powerful office—Udall, Watt, Babbitt? Two other writers, one from Wisconsin, one from Pennsylvania, who left their marks beyond the dry line—Leopold and Carson? The president, Reagan, who like Aspinall and Watt would wreck Stegner's West? You could add others, and quarrel with some I have named, but these names give a sense of the company. Surely, Stegner belongs in this list—near the top, it seems to me.

Even more telling will be Stegner's place in the years to come. The West was fresh and new for the first part of his life. But he told me once, "Now I see it growing old and worn." And it is. The essential hope is that in the Intermountain West, in the real Stegner country, the people will become ever more knowledgeable about the truths that he told. It is a measure of his greatness that there is a growing yearning for stability—for place in its fullest sense. There is a yearning, and increasing resolve, to make towns that a girl or boy can simply be children in; can feel, without necessarily thinking much about it, some healthy, open space; and can stay on a good while, free of the demons that have driven so many parents to push on toward the big find.

Such qualities may never thrive in the West. Today, as before, one would have to give the edge to the boomers and boosters, the mythmakers. Stegner never promised that anything would fundamentally change. He just told us what has happened, what went wrong, and what we might try.

But even if he left us no guarantees, at least he left us truths. Truths can blow away like cottonwood puffs in the wind or they can root and make forests—societies—out of barren ground. At the end of *Wolf Willow*, Stegner wrote about plain Whitemud, a nowhere out on the Saskatchewan plain. But he believed in Whitemud. It was his place. Perhaps, he thought, Whitemud would not only endure but, in time, thrive. "Give it," he wrote whimsically and lovingly, "a thousand years."[17]

Maybe it will take a thousand years. Maybe it will take a lot less. Maybe it will never take. It will depend on whether we keep to the truth. Now it's up to us.

PART I

STEGNER AS WRITER

STEGNER WAS FIRST AND FOREMOST—first and last—a writer. He was also a leading student, teacher, and critic of American literature, having personally walked the long trail of twentieth-century literary life virtually from its beginning to its end. His instincts as a realist kept him solidly grounded throughout; he did not stand on convention, but neither (as James Hepworth notes) did he experiment for experiment's sake. The essays in this section explore aspects of Stegner's record and character as a writer.

There are paradoxes to ponder here. Jackson J. Benson's essay stresses the personal nature of Stegner's literary output. "What made him stand out from other writers of his time," Benson writes, "was that his life and his work were so much of one piece." Benson examines the personal qualities—Stegner's childhood experiences, his raw talent, his work ethic, his breadth of knowledge—that so distinctively shaped the writing. Benson notes too Stegner's honesty, sometimes harsh but finely honed and nuanced in the narrators of his later novels. Through his narrators Stegner went about "the job of getting behind the myths, behind pretentiousness and deception, to find the truth. He was the ultimate realist."

John Daniel points out, however, that although one may find "lyrical passages of exquisite grace" in Stegner's nonfiction, one "will not find personal rhapsody without a context of geography and history. For Wallace

Stegner, the merely personal present is not the stuff of which literature is made." For Stegner the personal (like the West itself) could not be understood in isolation or fulfill itself in ignorance of its surroundings. "Stegner's project [was] to understand himself in the context of his world," Daniel explains. This task entailed a concern for the concentric circles in which the individual is embedded. "The wholest writers," according to Daniel, "are those with a complex sense of responsibility to nature, history, community, culture—to values that transcend their private epiphanies and miseries, to whatever it is that holds them in its sights and demands the most of them."

Recent literary criticism has highlighted the importance of race, class, ethnicity, and gender in understanding the sources and impact of literary expression. Stegner dealt with such subjects more implicitly than explicitly. Yet there is a deep vein of cultural criticism in almost everything Stegner wrote, and his treatment of cultural tensions, social equity, and gender roles emerges through his fictional characters and his nonfiction subjects.[1] Importantly, Stegner often addressed these themes within a broader environmental context (as, for example, in his novels *The Big Rock Candy Mountain*, *All the Little Live Things*, *The Spectator Bird*, and *Angle of Repose*). At a time when few writers were drawing such connections, Stegner was doing so subtly and attentively.

Another paradox, then: although Stegner anticipated such current themes, he did not partake of modern trends in literary style, technique, or content, and was especially averse to sensationalism. Melody Graulich's essay helps to resolve this paradox. Graulich examines the ways in which Stegner, in both his fiction and nonfiction, continually returned to one strong motif from his own life experience: protection. "Protection," Graulich writes, "is at the center of a web of Stegner's crisscrossing recurrent concerns: preservation, sanctuary, exposure, self-revelation, vulnerability, wounds, loss, safety, judgment, responsibility." The theme unifies Stegner's genres and connects his work as a writer to his work in other arenas. "Writing from the protective impulse," Graulich observes, "Stegner offers a momentary stay against modernist despair."

James R. Hepworth picks up another aspect of this thesis. Hepworth explores the positive role of the *wild* as a unifying theme in Stegner's writing. "Although it has yet to be properly documented," Hepworth writes, "one of Stegner's greatest contributions to twentieth-century letters is his reassertion of the Native American paradigm of nature as familial and relative rather than alien or foreign." A "good animal" (Stegner's term) will protect what is "familial and relative"; it will not willfully destroy its habitat. Stegner's fictional characters struggle with the conflict between their biological imperative and their social settings, while many of his histories and essays explore the consequences of this conflict in the world at large. Hepworth concludes that "our American idea of wilderness and the wild, however, has yet to enlarge itself to encompass Wallace Stegner's vision of them—or even the truth about ourselves."

There is room, of course, for criticism of Stegner's approach to these themes. (Hepworth's discussion of the evolving relationship between Stegner and Gary Snyder, another leading voice for the wild, is especially useful here.) Yet by carefully nesting his themes—of the personal within the sphere of human relationships, of social relationships within the culture, of culture within place, of all within human and natural history—Stegner found a satisfactory means of negotiating the tensions within and between contemporary literature and society.

CHAPTER TWO

WRITING AS THE EXPRESSION OF BELIEF

JACKSON J. BENSON

On a number of occasions, Wallace Stegner said that fiction was "dramatized belief." His former student and friend, Nancy Packer, has observed that "some writers are only interested in the aesthetics of writing, but for Wally, writing is a moral act. He deeply cares about what it means to live in a complex world."[1] He was a man of integrity, and what he wrote, both fiction and nonfiction, reflected that integrity.

Stegner stood out among other writers of his time in the degree to which his life and his work were of one piece. In life he was unpretentious, observant, and always, even in old age, learning. Much of his writing he devoted to the job of getting behind the myths, behind pretentiousness and deception, to find the truth. He was the ultimate realist. "In fiction," he wrote, "we should have no agenda except to try to be truthful."[2] His search was uncompromising.

As his former student James Houston has noted, Stegner never resorted to cheap tricks to make his writing popular or to publicize his work.[3] And to paraphrase what his former student Robert Stone said about one of his characters in a recent novel, Wallace Stegner believed in all those things we used to believe in—enduring love, friendship, generosity, kindness, fairness, duty, and sacrifice.[4] These were the things he wrote about in an age when they have so often been debased, discounted, or ignored—or even mocked as old-fashioned or irrelevant. He was one of a kind, a voice in the wilderness, a writer that we can treasure if we value the same things he valued.

He had no truck with talk about behavior. He would talk only about conduct. He insisted there was a difference. While he was a man of compassion, a liberal in the best sense of the word, he could not abide self-pity or a preoccupation with blame. He had been there himself. He had been the victim, he had been abused, and he had felt put upon, but came to reject that mode of thinking in his own life and despise it in others. "Only by repudiating the buck-passing irresponsibility of our period," he wrote, "can a man assert anything like his full dignity as an individual."[5] No contemporary writer has been less sentimental about those things that incline us to be sentimental: love, kindness, charity, forgiveness. Stegner has said about one of his mentors, Robert Frost, that "the real jolt and force of Frost's love of life comes from the fact that it is cold at the root. The same would seem to be true of the fiction writer.[6]

Stegner's realism is sometimes so bitter that it is hard for the reader to swallow it. Just as he was hard on himself, he was hard on his fictional creations. For him life was a test of character. When one reads *The Big Rock Candy Mountain* or *Angle of Repose*, one has the definite feeling, over and over again, that something bad is going to happen and, sure enough, again and again it does. That makes it hard to pursue a long novel with enthusiasm. He admitted this quality in his work: "I don't think I'm a particularly jovial or genial writer. I think a lot of my books are glum, bleak even."[7] He attributed this bleakness to his own habit of mind, which tended to be skeptical and pessimistic. Like Mark Twain he was a disillusioned romantic who found some merriment in human foibles, but in his heart he was constantly disappointed by human behavior—its intolerance, its self-centeredness, its self-deceptions, its rapacious misuse of the land. Like Twain, as he grew older he became increasingly pessimistic about humanity and its future—or as he put it, "I walk behind the times muttering about the way things are going."[8] As a realist he once admitted he probably would have been better suited to have lived and written in the nineteenth century along with Twain and Henry James.

He also admitted a distaste for much, but not all, contemporary writing: "I think now books are unfortunately pretty well riddled with what used to have a shock value, which means it has to have a higher voltage to shock now. . . . It's kind of a disease: attempting to be clever, sexy, or violent. It's a way of showing off."[9] When asked by the writer Sean O'Faolain how he managed to avoid being trendy, Stegner replied: "I write the kind of novel I can write. I don't know how to write out of the headlines."[10] When an interviewer in effect accused him of a lack of originality and innovation and cited *Angle of Repose* as an "extremely traditional novel," he replied that that label didn't bother him one bit: "I don't really aspire to write a novel which can be read backwards as well as forward, which turns chronology on its head and has no continuity and no narrative, which, in effect, tries to create a novel by throwing all the pieces in the bag and shaking the bag. . . . If you have to do that to be original, then I don't care about being original."[11]

Not only did he share aspects of Mark Twain's temperament but his career and his writing had much in common with the master, Henry James. Like James he had a major success in midcareer. For James it was *Portrait of a Lady*; for Stegner, *The Big Rock Candy Mountain*, which has taken its place among the classic novels of western experience. But both writers got better as they got older, James attaining the pinnacle of his powers with his late novels, Stegner reaching toward greatness with his last four: *All the Little Live Things*, *Angle of Repose*, *Spectator Bird*, and *Crossing to Safety*. Both writers wrote novels with very little plot; they tended to build dramatic situations that are realized internally rather than depending on dramatic actions. Their fictions were less, in Stegner's words, "a complication resolved than what Henry James was to call a 'situation revealed.'"[12] Both writers emphasized nuance of thought and emotion. Both were masters, too, of the indeterminate ending—the kind of ending that haunts you and makes you think back on everything that led up to it.

Who could ever forget the conclusion of James' *The Wings of the Dove*? Densher and Kate develop a cynical plan to get the dying Milly Theale's money by having her fall in love with Densher. But when Densher, against his own will, becomes very fond of her, the game changes. After Milly dies and Densher receives a legacy from her, which he declines, he turns to Kate and assures her that they will be as they were. "As we were?" she asks. "As we were," he replies.[13] But the reader shares their bitter knowledge that fortune has turned the tables on them and that their relationship can never be the same.

Nor can we forget the ending of *Angle of Repose*, where Lyman Ward lies in bed thinking, listening to the trucks laboring up the long grade on the highway near his Grass Valley home, wondering if he could possibly send for and thus forgive his unfaithful wife:

> Wisdom, I said oh so glibly the other day . . . is knowing what you have to accept. In this not-quite-quiet darkness, while the diesel breaks its heart more and more faintly on the mountain grade, I lie wondering if I am man enough to be a bigger man than my grandfather.[14]

Can the human heart ever be restored? Can things ever be as they were?

Both writers were very concerned with point of view in their fiction. It became a priority for Stegner only slowly, after considerable thought and some trial and error. In one of his interviews with Richard Etulain he said that "the more I study the art of fiction the more I think [point of view] is important."[15] In class he used the image of a garden hose running with a nozzle and without a nozzle—"You get more force," he would tell his students, "with a nozzle." Over the years, that nozzle would change. He began with a strictly limited, Jamesian point of view—that variation of the third person we call "center of consciousness." He admired the way that James, over the course of his career, limited his point of view more and more, so that as "he forced his story through smaller and smaller outlets . . . it acquired a special concentration and force."[16] But Stegner discovered there were subtleties in point of view "that Henry himself didn't know about. You don't have to be as rigid as he, and yet point of view is just as important, maybe even more important than he thought it was."[17] Stegner achieved his own greatness by turning away from that model to the first-person narrative, something that James had never used. It gave him many new advantages, but it also held some dangers.

With Joe Allston, first in "Field Guide to Western Birds" and then in *All the Little Live Things*, followed in the last novels by his first-person counterparts, Lyman Ward and Larry Morgan, Stegner invented the grumpy old man, his humor and pathos, long before Jack Lemmon and Walter Matthau discovered him. When asked by an interviewer why he used Joe Allston, Stegner replied half-jokingly, "I liked the chance to be crabby if I felt like being crabby and put it inside somebody else's mouth."[18] But the first-person perspective also had many technical advantages: it "encourages you to syncopate time . . . [and] allows you to drop back and forth, almost at will, freely."[19] Interpenetrating the past and the present, which he does frequently in these final novels, was important to him as a writer who was also a historian, someone who became a historian because he felt so ignorant about his own past. He admired Faulkner for the richness of his associations from the past and mourned the poverty of these associations in most western novels.

Stegner found that the first-person narrative brought him closer to his work and gave him freedom. It allowed him to be pretty much himself, or pretty much what he might imagine himself doing or saying, while speaking through a mask the way Conrad speaks through Marlowe. That is to say, it allowed him to dramatize his beliefs. Yet Stegner had avoided that technique for years because he was afraid it might lead him to be windy and he was afraid—as a very private man—of inadvertently revealing too much of himself. His fear was justified. However mightily he tried to persuade his audience that although he contributed much to his first-person narrators he was not identical to them, the reader's identification of author with speaker was almost inevitable.

Stegner treated this confusion of writer and character with humor at times, but sometimes with irritation. On the book tour for *Crossing to Safety,* he found it amusing that so many in his audiences were surprised to see that his wife Mary was not wearing braces. On the other hand, he'd get annoyed with me as his biographer when I'd ask about models for his characters or the factual background of his works—the sort of thing that readers of biography are most interested in—though he would sometimes talk quite freely in public (as he had occasionally written freely) about sources of material in his own life and opinions that he and his narrators shared.

Regardless of his occasional protests, it is clear that the man and his work were one in any number of ways. Most of his work, of course, is autobiographical. At times he would admit this but emphasize that a writer's life is the only material he has to draw on. It may have been his life, he admitted, but the material of that life was now given form and direction. It didn't matter how autobiographical his work was: "Both fiction and autobiography attempt to impose order on the only life the writer really knows, his own."[20] The point for us is to recognize how closely his life and work were joined: it isn't a matter of the narrator's identity, it's a question of his values and beliefs.

The metaphor he used for this artistic process was the lens. In an essay on his philosophy of writing, he wrote:

> One page or six hundred, a fiction is more than a well-carpentered entertainment. It is also more than the mirror in the roadway that Stendhal said it was. Because a good writer is not really a mirror; he is a lens. One mirror is like another, a mechanical reflector, but a lens may be anything from what is in your Instamatic to what makes you handle your Hasselblad with reverence. Ultimately there is no escaping the fact that fiction is only as good as its maker. It sees only with the clarity that he is capable of, and it perpetuates his astigmatisms.[21]

What, then, were the properties of this lens and the characteristics of its images? What were the aspects of Wallace Stegner that produced the fiction which dramatized his beliefs and whose qualities his readers admire? First, certain unique experiences while growing up; second, an enormous talent; third, an incredible self-discipline; and fourth, a lifelong thirst for learning that resulted in an unusual breadth of knowledge.

Alone among the major writers of his time, Stegner's life spanned the years from the horse and buggy to the information age, giving him an unusually long perspective. Growing up on a farm on the last homestead frontier gave him an early immersion in nature and an intuitive grasp of the importance of humanity's relationship to the earth. As well, it planted the seed for his appreciation of a sense of place as a basis for fiction. Stegner has repeatedly confirmed the importance of that appreciation: "I don't think you can be a decent novelist without having a sense of place. I know some novelists who don't have it, and it seems to me that without knowing it, it's like a vitamin deficiency."[22]

Mocked for his smallness and sickliness as a child and abused by a brutish father, he learned to stick things out and not complain. He grew up to be a gentle man, but also a very tough man, with no self-pity and not a sentimental bone in his body. Alone much of the time as a child on the farm and then isolated as a teenager because of his father's dubious career as a rumrunner, he found his comfort in books, his only approval coming from his mother and teachers for his schoolwork. He became a voracious reader and an outstanding student throughout his school and college years. He resented his father's rugged individualism and pursuit of the get-rich-quick western dream. In reaction he supported throughout his writing career the importance of cooperation and spirit of community. Family and friends were essential to him and among his primary subjects.

Stegner's writing talent was far greater than generally appreciated. Early on he learned to love literature and the sound of words. He had a remarkable gift for remembering poetry and verse, and that gift in turn led to the development of a discriminating ear for the rhythms of prose. Those who knew him well have testified that metaphor was a habit of mind which came out even in ordinary conversation. He had what he himself, in talking about others, called "sensibility." In an essay he wrote for *Saturday Review* about judging student writers, he declared that sensibility was essential for a writer, and even "though [it] . . . can be trained and sharpened, it cannot be created or acquired." Rather than being one of those Henry James figures on whom nothing is lost, Stegner said he "should like to be one who seizes from everything some vivid impression," for "there ought to be a poet submerged in every novelist."[23]

On sheer talent alone, without knowing much if anything about storytelling technique, he sat down and wrote his first long manuscript and won a major prize for it. Perhaps the best testimony to his talent would be to reread the opening paragraph of one of his earliest short stories, "Bugle Song," which he wrote in 1937 during his first year teaching at the University of Wisconsin. I mention the early date of its composition—Stegner considered it his first professional story—to emphasize the native talent of the writer that produced it. This paragraph illustrates that no one has been better able than Stegner, even at the beginning of his career, to create an atmosphere, a mood, or to suggest a condition of the soul by providing a suggestive landscape:

> There had been a wind during the night, and all the loneliness of the world had swept up out of the southwest. The boy had heard it wailing through the screens of the sleeping porch where he lay, and he had heard the washtub bang loose from the outside wall and roll down toward the coulee, and the slam of the screen doors, and his mother's padding feet after she rose to fasten things down. Through one half-open eye he had peered up from his pillow to see the moon skimming windily in a luminous sky; in his mind he had seen the prairie outside with its woolly grass and cactus white under the moon, and the wind, whining across that endless oceanic land, sang in the screens, and sang him back to sleep.[24]

Indirectly, metaphorically, the paragraph introduces the thematic conflicts of the story: untamed nature and frontier values versus domesticity and the values of civilization. But beyond this what strikes us most is its haunting beauty, which comes out of the paragraph's imagery and its nearly perfect rhythmic pattern. What we have here is part and parcel of what he was: his childhood experience and sense of place, his metaphorical imagination, and his sensitivity to language.

Talent is important. But just as important is the self-discipline with which that talent is developed and applied. Every schoolday while he was teaching at Wisconsin, Stegner had a ten o'clock class. Since he had already decided he would be a writer as well as a teacher, he would write every morning starting at seven and then rush off up the hill to class a little before ten. Thus began a routine he would follow the rest of his life: sitting down to write early every morning and, when he was not teaching, into the early afternoon. Whether the writing was going well or not going at all, he stuck with it, not letting himself get up from his chair until his period was over. This for him was the basis for being a professional—staying with it. By the time he went to Harvard, after Wisconsin had refused to promote him, his wife Mary was worried much of the time about his health because he worked such long hours and spent so little time resting.

I had some personal experience with his self-discipline. For more than seven years I spent several weeks off and on during the year with him in his home office, first interviewing him and then going over his papers. I found myself impressed, not only with his energy, but with his sense of purpose. I could be sitting at a table in his office early in the morning reading through some letters. After asking if I were comfortable, he would turn to his old upright typewriter and stay with it for hours. While I would take numerous breaks to walk in the yard or go to the bathroom, he would still be banging away. He had incredible powers of concentration. Unlike other notable figures I had run into over my years of interviewing for biographies, Stegner always had his phone number listed in the directory and the phone would ring now and again all morning. If Mary was home, she would screen the calls, passing on those she thought important. Some calls had to do with ongoing projects, but many were requests of one sort or another. He was always patient and heard the caller out. Usually he then declined the offer to speak or to write something—often an introduction to someone else's book. (In exasperation after one such request, he turned to me and claimed to be "the champion introduction writer of all time.") But after the call, he would swing back to the typewriter, not skipping a beat. Several days after Stegner died of injuries in a car accident at the age of eighty-four, Mary went out to clean his office and put things in order. Over his desk, pinned to the wall, she found a handwritten list of ten items he had been planning to write in the weeks following his trip to Albuquerque: articles, speeches, and introductions.

Having interviewed hundreds of people, many of them prominent in their fields, for biographies of Steinbeck and Stegner, and having worked with academics for several decades, I think I can testify with some authority as to Stegner's intellect. He was simply, by far, the brightest man I've ever known. And although my description here has emphasized his strict moral standards and stern self-discipline, I should mention that like most bright people, he had a wonderfully witty and playful sense of humor. What complemented his intelligence and made it even more impressive was an incredible memory, a vast storehouse of general knowledge. He was interested in many things, from geology to sports, from Native American history to classical music, from the cooperative movement to conservation. He was a reader, a learner, all his life. And whether he was at graduate school at Iowa or teaching at Utah, Wisconsin, Harvard, or Stanford, he always thought that he was behind, that his colleagues knew much more than he and that he would have to work like hell to get caught up. And he did—worked like hell and got caught up.

There is no doubt that his superb intellectual equipment was in large part responsible for the depth and range of his accomplishments. He had ground and polished a lens, applying the grit of experience and self-discipline to the raw materials of talent and intelligence to produce images, a body of work, of more consistent quality and greater breadth, perhaps, than any other writer of his time. In addition to receiving awards for his novels, histories, and short stories, he was presented later in his career with a half-dozen lifetime achievement awards. In his case, to a large extent, the man was his work—a lifetime spent in honesty, realism, and dedication to the truth. He was a good deal more like the stereotypical western hero than he ever would have admitted. He was one tough customer. And notice, I smile when I say that.

CHAPTER THREE

WALLACE STEGNER'S HUNGER FOR WHOLENESS

JOHN DANIEL

ONE NIGHT LAST WINTER as I was rereading some of the essays in *The Sound of Mountain Water*, I heard myself say to my wife, "Wallace Stegner may be our *wholest* American writer." As it was two in the morning and she was asleep, my wife did not reply. I took her rapt silence as affirmation and decided I would try to track down what my remark might mean.

I can't think of another writer who has distinguished himself or herself in so many fields of prose: the novel, the short story, the essay, memoir, history, biography, literary journalism. Only poetry, it seems, eluded Stegner's pen. He liked verse, though, and when I was around him he frequently quoted lines and considerable passages from Milton, Wordsworth, or Robert Frost, some of which he had carried in memory all the way from his early schooling in the 1910s and 1920s. He liked some contemporary poetry, too, particularly that of Wendell Berry and William Stafford. Once, giddy from having my first essay accepted by *Wilderness* magazine, I told Stegner that I might quit writing poems altogether. "Don't you dare," he said. "It was the poets who led us out of the caves."

Imagine my pleasure, then, when I discovered that Stegner himself had scribbled a verse or two. One day in the 1980s at Stanford University, when I should have been reading freshman English papers, I wandered instead into an unoccupied neighboring office and browsed through a few dusty books. Opening a little volume entitled *Historia de España*, copyright 1923, I was shocked to find on the flyleaf the signature of Wallace Stegner, along with "Español 5" and "Universidad de Utah." I leafed ahead and found page after page of penciled Stegner artwork (labeled "futuristic" by the artist) and various marginalia, including two short poems. One of them was a kind of biblical haiku:

> Yea verily I say unto you
> ¿How's your old lady?

The other delivers itself in five snappy lines:

> Socrates
> Apple crates
> Insurance rates
> Pair of eights
> Soul mates

More lively than some poetry of the 1920s.

I can't resist mentioning that on another page I caught young Stegner practicing the "S" in his last name, shaping it something like the G-clef and something like a dollar sign. He sensed, perhaps, that in the life ahead of him he might need to be signing his name a lot. I also found three pairs of seeming nonsense words that suddenly came clear as female names written backwards. Whatever Grace Anderson, Nola Cook, and Carol Barclay saw in young Stegner, it seems likely that young Stegner saw poetry in them.

BUT THE WHOLENESS of Wallace Stegner I value most is not a matter of poetry and not only a matter of his versatility. It has to do with what he put into his prose and put into himself. Listen to the beginning of "Overture: The Sound of Mountain Water," his shortest essay and to my mind one of his best:

> I discovered mountain rivers late, for I was a prairie child, and knew only flatland and dryland until we toured the Yellowstone country in 1920, loaded with all the camp beds, auto tents, grub-boxes, and auxiliary water and gas cans that 1920 thought necessary. Our road be-

> tween Great Falls, Montana, and Salt Lake City was the rutted track that is now Highway 89. Beside a marvelous torrent, one of the first I ever saw, we camped several days. That was Henry's Fork of the Snake.
>
> I didn't know that it rose on the west side of Targhee Pass and flowed barely a hundred miles, through two Idaho counties, before joining the Snake near Rexburg; or that in 1810 Andrew Henry built on its bank near modern St. Anthony the first American post west of the continental divide. The divide itself meant nothing to me. My imagination was not stretched by the wonder of the parted waters, the Yellowstone rising only a few miles eastward to flow out toward the Missouri, the Mississippi, the Gulf, while this bright pounding stream was starting through its thousand miles of canyons to the Columbia and the Pacific.
>
> All I knew was that it was pure delight to be where the land lifted in peaks and plunged in canyons, and to sniff air thin, spray-cooled, full of pine and spruce smells, and to be so close-seeming to the improbable indigo sky. I gave my heart to the mountains the minute I stood beside this river with its spray in my face and watched it thunder into foam, smooth to green glass over sunken rocks, shatter to foam again.[1]

The piece goes on for two more paragraphs of river description and contemplation, paragraphs as lyrical as any Wallace Stegner wrote. If I had camped by that river and written a short essay about it, I wouldn't have included the preliminary details of the river's history because I wouldn't have known them and wouldn't have looked them up. I might have mentioned the Continental Divide but probably not the source and length and terminus of the river—again, I wouldn't have known or sought that information. In short, I would have written, whether in prose or in verse, a lyric poem about my first encounter with a mountain river. And maybe that suggests one reason why Wallace Stegner didn't write verse. For the poet—if I may overgeneralize to make a point—the lyric moment is consuming and aesthetically sufficient unto itself. For Stegner, the lyric moment may be intense but is never in itself aesthetically sufficient. In his nonfiction you will find lyrical passages of exquisite grace, but you will not find personal rhapsody without a context of geography and history. For Wallace Stegner, the merely personal present is not the stuff of which literature is made.

Listen to the rhetorical strategy by which he informs you of what, as an adult, he has informed himself. The information is carried in a string of negative expressions, a technique he used frequently. I call it the informational negative:

> I didn't know that it rose on the west side of Targhee Pass . . .
>
> The divide itself meant nothing to me. My imagination was not stretched by the wonder of the parted waters . . .
>
> All I knew was that it was pure delight . . .

That "All I knew" works two ways: a full cup to the enthralled eleven-year-old, but only half a cup to the adult remembering and writing the experience. And only half a cup, I think that writer would argue, to anyone who would know the American land in its wholeness.

LET ME ENLARGE MY POINT by turning to one of Wallace Stegner's masterpieces. *Wolf Willow* is an account of return and remembering, a midlife putting together of the place and culture he came from. He announces the project of the book on the first page: "That block of country between the Milk River and the main line of the Canadian Pacific, and between approximately the Saskatchewan-Alberta line and Wood Mountain, is what this book is about. It is the place where I spent my childhood. It is also the place where the Plains, as an ecology, as a native Indian culture, and as a process of white settlement, came to their climax and their end."[2]

The personal reference is one short sentence between two long ones. Autobiography, Stegner is signaling the reader, is only one element in a much larger story. The first few pages of the book are mostly history, first human then natural; he shows us the past he didn't come from—the gun-toting culture of horse-opera movies—and the immense Plains landscape that did indeed form his soul and character. The present-time narrative of his return to that country as a man in his forties begins unobtrusively. Not until page six does he "poke the car tentatively eastward . . . from Medicine Hat . . . easing watchfully back into the past." As he homes in on the town of Whitemud, the house and play-haunts of his childhood, he sizes memory against present perception, and the personal narrative expands. He is literally remembering, piecing together his past, making it whole again. The process culminates, not as he views the house his father built or the businesses on Main Street, but when he identifies a pervasive smell that captures him by the Whitemud River. It is wolf willow, a native riparian bush. He makes his past whole by exercising our oldest, most deeply rooted sense:

> For a few minutes, with a handful of leaves to my nose, I look across at the clay bank and the hills beyond where the river loops back on itself, enclosing the old sports and picnic ground, and the present and all the years between are shed like a boy's clothes dumped on the bathhouse bench. . . . A contact has been made. . . . A hunger is satisfied. The sensuous little savage that I once was is still intact inside me. [p. 19]

Coming to that realization would have constituted the bulk and main thrust of the book for many writers: I have returned, I have remembered who I was and who I am. But for Stegner the personal is not sufficient. This realization comes in chapter one, the first of nineteen plus an epilogue. The next two chapters amplify the personal story, but in the following nine he progresses to the larger weave in which his family's life is only one thread. He gives the history of his town and region, the history that he did not receive as a child, because it had not been recorded or thought relevant and because the white-settlement phase of it was only beginning in 1914 when his family arrived. The middle-aged writer is regretful, even resentful, over what he missed as a boy. In chapter nine the informational negative introduces paragraph after paragraph, ringing like a litany:

> The very richness of that past as I discover it now makes me irritable to have been cheated of it then. [p. 112]

> All of it was legitimately mine, I walked that earth, but none of it was known to me. [p. 112]

> We were not informed in school that the graces of imported civilization first appeared here at Fort Walsh. [p. 113]

> I wish we had known it. I wish we had heard of the coming of the Sioux, when they rode northward after annihilating Custer's five troops on the Little Big Horn, a whole nation moving north, driving the buffalo before them, and with the soldiers from every army post between Canada and Texas on their track. [p. 114]

> I knew the swallows and muskrats, and was at ease with them. . . . But Time, which man invented, I did not know. I was an unpeopled and unhistoried wilderness, I possessed hardly any of the associations with which human tradition defines and enriches itself. [p. 122]

In an interview recorded in 1987, Stegner likened the absence of those associations to a deficiency disease, a condition that stunts the human spirit as malnutrition stunts the body. But he refused to remain stunted. To heal means to make whole, and Wallace Stegner, deprived of wholeness as a child, later followed his hunger and healed himself in a singular and prodigious way. He made himself into the Herodotus of the Cypress Hills, unearthing and imagining and writing the history he hadn't known.

Stegner said that the main task of *Wolf Willow* was history: autobiography and fiction were adjuncts to the historical account. I have a different view. I see the book as shaped and powered by the needs of memory seeking its wholeness. Personal memory comes to completion in the first three chapters. But personal memory, as the author remarks several times, is uneven and unreliable, and even in its wholeness is insufficient. And so he augments personal memory with the collective memory of history. But even that larger memory is limited and insufficient—insufficient, at least, to a writer of Wallace Stegner's hunger. And so, with the epic novella "Genesis" and its sequel story "Carrion Spring," he takes a further step. He extends his act of memory beyond the horizon of personal and historical fact into the greater realization of fictional creation, and a book that would have been ample as a memoir and history becomes something more, the very most its author could make it—a full-immersion baptism in the times and landscape that made him, one of the great acts of knowing in the American literature of place.

"I thought I could get more truth into a slightly fictionized story of the winter that killed the cattle industry on the northern plains than I could into any summary," Stegner remarked in his essay "On the Writing of History."[3] His comment characterizes his entire project as a writer. As he worked up a ten-thousand-word autobiography for a reference book near the end of his life, he observed that he had the feeling he had written it all before. "Most fiction writers have already written their autobiographies piecemeal," he stated, "overtly or covertly, and go on doing it every working day."[4] The Latin root of "fiction" is the verb *fingere*, and "to feign" is only one of its meanings. It also means to mold or to form. Memory gives form to the flux of subjective experience, history gives form to collective experience held in memory, and fictional narrative further forms memory and history into the wholer truth of art.

Without "Genesis," a reader of *Wolf Willow* would know *about* the harsh climate of the Canadian plains; with it, he knows that brutal cold in his bones and blood. With "Carrion Spring" he knows from a woman's perspective the isolated home imprisonment of that same awful winter. Both stories are fundamentally about cooperation: a band of buckaroos on roundup suddenly must work together to survive; a newly wedded husband and wife struggle to settle their claims on place, culture, and each other. Both reflect another form of wholeness Wallace Stegner hungered for—the wholeness of community. By and large, acts of individual heroism did not interest him so much as acts of individuals working and neighboring and loving together.

That was the wholeness he wished for his mother, who had the character and skills to be a sticker, a community builder, but was married to a man who did not. "How hungry you were!" he addresses her in "Letter, Much Too Late." "How you would have responded to the opportunities ignored by so many who have them!"[5] And in these words, of course, his own hunger as a boy is spoken, the hunger that in *Wolf Willow* gave him a recurrent bolt of joy at the sight of town, "looped in its green coils of river, snug and protected in its sanctuary valley," when the family trekked back from the wheat farm in the fall" (p. 11). Town was school and games and friendship, the post office, travelers from afar. Summers on the homestead—three hundred and twenty acres of wind and grass bounded by one iron post and three survey stakes—his hunger for human association could express itself only in a passionate regard for the footpaths the family walked into the prairie in their daily living. In the last chapter of *Wolf Willow*, titled "The Making of Paths," Stegner tells of his frustration when other members of the family cut across to the privy from the wrong corner of the house, ignoring the proper trail. "I scuffed and kicked at clods and persistent grass clumps," he writes, "and twisted my weight on incipient weeds and flowers, willing that the trail around the inside of our pasture should be beaten dusty and plain, a worn border to our inheritance" (p. 272).

When three summers of drought did in his father's farming fantasy and set the family on its erratic way again, the eleven-year-old Stegner (soon to see his first mountain river and his first flush toilet) took with him hungers he didn't know he had, hungers of incompleteness that would drive him through a mighty career. But he did leave with one kind of completeness already realized. In six years he had come to know intimately the land and life and weather of the Plains. He knew the "pushing and shouldering wind, a thing you tighten into as a trout tightens into fast water . . . a grassy, clean, exciting wind, with the smell of distance in it, and in its search for whatever it is looking for it turns over every wheat blade and head, every pale primrose, even the ground-hugging grass." He knew the circling horizon always miles away, "as clean a line as the nearest fence." And he knew the immense sky, alive with "navies of cumuli, fair-weather clouds, their bottoms as even as if they had scraped themselves flat against the flat earth" (p. 7).

In a 1987 interview, Stegner said he could draw pictures of Plains wildflowers he hadn't seen since he was ten.[6] He knew the particulars of that country, and he knew its changefulness, the way tornadic storms could boil up in a blue-black sky. But he also knew, beyond its things and motions, the country's elemental permanence—disk of earth and bowl of sky, exact circle of horizon, segmented lines of fence and roadway. "A country of geometry," he calls it in *Wolf Willow*, a "Euclidean perfection" (p. 7). It is the paradox of that landscape to be simultaneously empty and archetypally whole, and human presence there is paradoxical as well—a person is on the one hand tiny, on the other "as sudden as an exclamation mark, as enigmatic as a question mark. . . . At noon the total sun pours on your single head; at sunrise or sunset you throw a shadow a hundred yards long" (p. 8).

Stegner argued in his essay "Ansel Adams and the Search for Perfection" that the Sierra Nevada, particularly Yosemite, had taught Adams how to see and made him the photographer he was.[7] It may be reasonable, also, to argue that the Northern Plains taught Wallace Stegner how to see and made him the writer he was. Certainly a writer under the influence of that whole and empty landscape is likely to look outward for his completeness—to look beyond himself in a way that writers, say, in the temperate jungle of the Northwest Coast or the skyscraper canyons of New York City might not. Such a writer is likely to know the singularity of things and creatures lit by a sun that shines in summer from four o'clock in the morning until nine at night. He will be interested in community, a student of how people get by. He is likely to carry within him a sense of possibility as immense as the land around him. And he may want to make a mark—a question mark, an exclamation mark, a mark perhaps like the pathways that fired his young mind. It will be a seemly mark, but not a small one. An artist whose mind comes to light in that expansive space will be no miniaturist.

"I may not know who I am," Stegner wrote in *Wolf Willow*, "but I know where I am from" (p. 23). That famous comment may be a bit misleading. Wallace Stegner knew very well who he was—knew himself better than anyone in my experience—and he knew who he was exactly because he knew where he was from. In one of his last essays he writes, "There is something about exposure to that big country that not only tells an individual how small he is, but steadily tells him *who* he is. I have never understood identity problems." And he goes on to write: "I knew well enough who, or *what,* I was, even if I didn't matter. As surely as any pullet in the yard, I was a target, and I had better respect what had me in its sights."[8]

An interesting statement. The pullet in the yard had been the target of a ferruginous hawk dropping out of an empty sky just a few yards from young Stegner, an event he witnessed several times. But what in that immensity had him in its sights? His sense of vulnerability may have had something to do with being a sickly child whose mind was two years ahead of his body in school, and who grew up, in the adult writer's words, "hating my weakness and cowardice" (*Wolf Willow*, p. 131). Lashed to a survey stake in a one-foot hole in the prairie, waiting out a cyclone, may have contributed as well. But in the 1987 interview he put the sense of being a target somewhat differently. He spoke of the uncanny doubleness one feels beneath that enormous sky, a sense "of observing everything else the way God may be observing you."[9]

Wallace Stegner didn't refer to God very often. I don't know what God meant to him, whether sky or wind or stars, biological life, all of those, or something transcending them. But I suspect that the sense of being observed would come easily in the Plains country, and with it a sense of being inevitably known, of being a question mark unable to hide, and I suspect this sense contributed considerably to the healthy hunger of Wallace Stegner's eighty-four years. A targeted man, a man in the sights of a power he respects and fears and loves, is likely to work as hard and as well and as long as he can.

Talking once with a group of Greek writers about having grown up without history, Stegner was surprised to hear from some of the Greeks that they envied him. Their rich and lengthy past felt like no blessing, they said, in the diminished present of their culture. "But I envied them more than they envied me," Stegner wrote, "for what they had was what I had spent my life hopelessly trying to acquire."[10]

Hopelessly? A targeted man might look at his life that way, I suppose, but the statement is hopelessly inaccurate. Stegner did indeed grow up, as he puts it in *Wolf Willow*, in a "dung-heeled sagebrush town on the disappearing edge of nowhere, utterly without painting, without sculpture, without architecture, almost without music or theater, without conversation or languages or bookstores, almost without books" (p. 24). He was indeed "charged with getting in a single lifetime, from scratch, what some people inherit as naturally as they breathe air" (p. 2). And get it he did. There was nothing hopeless or halfway about the lifelong acquisition of wholeness he embarked upon once his hunger took hold of him. In his senior year of high school he shot up six inches, finding himself suddenly as big as his classmates. And in his inner life, his cultivation of himself, you get the feeling he took off at the same rate and scarcely slowed.

Hunger is a powerful need, and it can be an asset. Ask any writer, any artist, if he does his best work after a big meal or before. Ask any mountain lion. Wallace Stegner knew the value of learning and training and opportunities for the soul's expansion, and sought them as energetically as he did, precisely because they had been unavailable to him for much of his youth. Few writers now, I would hazard to say, grow up deprived in that way. For American writers at the end of the century I would say that the situation is directly opposite to Wallace Stegner's situation at the century's beginning. These days we come up well schooled, richly provided with books and museums and performance halls, hundreds of college creative writing programs waiting to nurture and unfold our talents.

And what has this meant for our literature? It's burgeoning, certainly. A great plenty is being written and published, and some of that plenty is good. I have to wonder, though, if contemporary literature isn't suffering in some ways from a lack of hunger in its creators. Where is the ambition, for instance, to be as knowledgeable of history and geography and other fields beyond the specialized craft and criticism of creative writing? Where is the desire to write well in several genres? Where, outside the field of nature writing per se, is the American landscape in our literature? Where is the awareness of tradition—of one's place in what Stegner called "the great community" of recorded human experience? And where, especially, is the writing that seeks its wholeness outward, recognizing that personal rapture or private torment is insufficient material for the making of literature? It was Wallace Stegner's project to understand himself in the context of his world. Increasingly, it seems, it is the project of contemporary writing to understand the world—or fail to understand it—in the context of the self.

No doubt I am exaggerating to make my point. Today's literature is by no means entirely self-referential or self-enclosed. And I certainly don't wish to glorify cultural deprivation and repackage and remarket the destructive myth that art, in order to be authentic, must be the product of hardship. Talent finds its way, whether born with a silver pen in its hand or with dung on its heels, and our time has its share of talent. But talent, even genius, does not make a writer. Neither does the capacity to emote. Literature is more than self-expression. "If a writer has only himself to say," Wally once told me with a mild smile, "his work will be kind of thin."

That is an old-fashioned idea, and it came from an old-fashioned man. "I really don't belong in the twentieth century," Stegner said in conversation with Richard Etulain. "My demands upon life are nineteenth-century demands."[11] Elsewhere he worried that "the principles of restraint, proportion, and a wide representation of all kinds of life—the principles I have tried to live and write by—have all been overtaken and overwhelmed."[12] His concern has greater justification now than when he first expressed it in the 1970s. As book publishing in America becomes a subsidiary business of corporate media conglomerates, the sensationalism of sex and violence that dominates film, television, and computer games increasingly drives the book market as well. Publishers have abdicated their traditional sense of cultural responsibility, paying huge advances for loud, flashy, and often poorly written novels and for mediocre nonfiction by or about celebrities. The more modest but well-crafted books they used to keep in print are now quickly remaindered if published at all. Creative writing is in danger of becoming another supply-and-demand commodity of the mass-culture marketplace.

And yet, for all that, a book as quiet and quirky as *Wolf Willow* is in print today and has hardly been out of print in the thirty-four years since its initial publication. *Crossing to Safety*, Stegner's late novel on the very unsensational subject of friendship, has sold thirty-five thousand hardback copies and countless paperbacks since it appeared in 1987. In fact, three and a half years after his death, practically every major book Wallace Stegner wrote is now in print in one edition or another. When I see him in my mind these days, he's smiling, and maybe not only with self-satisfaction. Maybe his own late-life and posthumous success would convince even him that there remains a sizable audience for old-fashioned books that bring into focus what is worthy and enduring in human experience. The wholest writers are those with a complex sense of responsibility to nature, history, community, culture—to values that transcend their private epiphanies and miseries, to whatever it is that holds them in its sights and demands the most of them. Wallace Stegner was that kind of writer. For sixty years, every morning till noon, he extended a carefully considered pathway out of the nineteenth century through the broad terrain of modern American life. That path, inconspicuous but clearly defined, democratic but demanding, is one of the routes most likely to lead us to a future we want to inhabit.

CHAPTER FOUR

RUMINATIONS ON STEGNER'S PROTECTIVE IMPULSE AND THE ART OF STORYTELLING

MELODY GRAULICH

IN WALLACE STEGNER'S *Crossing to Safety* (1987), the narrator, Larry Morgan, a fiction writer, muses about the nature of storytelling: "Drama," he says, "demands the reversal of expectation, but in such a way that the first surprise is followed by an immediate recognition of inevitability."[1] His "quiet" story of friendship, he goes on to say, will offer no such drama. Yet after a particularly intimate conversation with his friend Sid Lang, Larry finds that to his surprise he has begun to feel "a little protective" of his friend.[2] Larry may be surprised, but the longtime reader of Stegner's work immediately recognizes the inevitability of Larry's desire to protect Sid.

Often in search of their own "safe places," to quote Astrid from *The Spectator Bird*, the characters Stegner most admires, or those with whom he most closely identifies, are commonly aspiring "protectors" who finally fail in a host of ways to protect those they love. In *The Big Rock Candy Mountain*, Elsa Mason tries to protect her young son Bruce from his father's rages, but she can offer no real safety or sanctuary. The middle-aged Bruce Mason of *Recapitulation* tries to "preserve" control, to defend himself, by "inking out" his younger self; his "safe place" comes at great cost. In *Angle of Repose*, Oliver Ward's efforts to preserve his wife's role as one of the "protected" women of her time only reinforce her distance from western life and her eventual estrangement from him. Trying to protect his grandmother from Shelly's ahistorical judgments, Lyman Ward leaves her open to be judged in his own terms. Neither Joe Allston nor Larry Morgan can protect the women they love from life's darker designs. The theme of protection informs Stegner's political analysis, too, as we see John Wesley Powell's heroic but ultimately failed attempts to protect the West from the politicians, the conservationists' inability to preserve the wilderness.

Nor should we be surprised that the protective Larry Morgan is a writer. Although Stegner often posed as an "old-fashioned" realistic writer, many of his novels ruminate on the nature of storytelling, to borrow the subtitle from one of his last essays, "Ruminations on the Art of Fiction." Stegner's storytelling, his narrative stances, center on what I once considered "an ethic of protection." Now I'm not so sure "ethic" is quite the right word; nor is "theme" or "trope" or "issue" or "technique." I do know that protection is at the center of a web of Stegner's crisscrossing recurrent concerns: preservation, sanctuary, exposure, self-revelation, vulnerability, wounds, loss, safety, judgment, responsibility. In much of his work he assumes that our sanctuaries, literal or metaphorical, will be threatened, that wounds will occur, that someone must take responsibility to protect, and that someone must make sense of the failure or inability to protect. Stegner ruminated over a long career on a theme that emerged from his own life, turning the idea this way and that, thinking about it from a variety of angles.

Protection is at the center of Stegner's narrative choices. His narrators are often torn between a desire to expose the truth and a need to protect the characters they create. The desire to protect at once exposes and redresses feelings of separation, abandonment, and estrangement. Narrators cannot save characters from pain and loss, from stubbornness and self-revelation, from the past, from each other. Those who cannot be protected in life cannot be fully protected in art, either, but their stories can be told with compassion and respect. The protective impulse counters feelings of isolation, distance, separation. It makes us understand the basis of our humanity—as Stegner suggests in "A Capsule History of Conservation," where he defines himself primarily as a "preservationist" seeking to protect wildlife and wilderness. There he locates the basis of his protectionist ethic in an essay by Aldo Leopold in which Leopold "argues for the kind of responsibility in our land relations that civilized people are supposed to show in their human relations."[3]

The protective impulse surfaces in crucial scenes throughout Stegner's work: in stories with traditional western themes like "Carrion Spring" from *Wolf Willow*, where a ranchwoman, feeling powerless in the face of a brutal natural world, seeks to protect a coyote pup; in *The Spectator Bird*, Stegner's tribute to Hawthorne and James, where the American "innocent" Joe Allston would like to protect the Countess Astrid Wredel-Krarup from experiencing the corruptions of the "Old World"; in early stories like "Beyond the Glass Mountain," where the main character helplessly attempts to intervene in an old friend's failed life; and in late works like *Recapitulation*, where a game of strip poker turns into a struggle over exposure and protection. In this essay I want to look at Stegner ruminating about protection primarily in three novels, exploring how the theme emerges both in plots and narration. Let us begin by examining two crucial scenes, in the autobiographical novel *The Big Rock Candy Mountain*, that might help us see the genesis of the theme in his life experiences and early narrative choices.

IT IS ONE OF STEGNER'S most dramatic scenes: when the enraged Bo Mason rubs his young son's face in his own excrement. The scene focuses on Elsa, the mother, who "crowded Bruce behind her back protectively," who hit Bo with a piece of stove wood, fought him for her son, who assured Bruce afterward that "we won't let him do it any more." She then sheltered the four-year-old in her arms, trying to offer him a place "soothing, soft, safe."[4] In Stegner's fiction the "safe place" is always threatened. The mother can soothe the wounded child, but her decision to return to Bo assures that he will "do it"—abuse his son—many more times; as he buries his head in her lap again and again, she can offer only "muffled sanctuary" (p. 203). Years later when she expects him to work out his relationship with his father by himself, the point of view shifts to Bruce: "The boy felt her exodus like a surrender, a betrayal" (p. 198). The mother struggles to protect her fearful but defiant son, yet she cannot. Here and in later work, the son pays tribute to her efforts, not her successes.

This crucial scene echoes throughout the novel—indeed, I will argue, throughout Stegner's career. Stegner creates drama by leaving open the question of whether Bruce himself remembers the scene, told from Elsa's point of view, until late in the novel. But he has more significant reasons for telling the scene from Elsa's point of view. He evades judgment. In the subtle dynamics of the scene, he sympathizes with her feelings of powerlessness and guilt, those absolving her of blame from his point of view, yet causing her to blame herself, from her own. To return to the Leopold quote, the issue of human responsibility becomes murky and complex. Effort is more important than success. The authorial voice fails to protect Elsa from her own judgments, which acknowledge her failure and her responsibility, but it does protect her from his own.

Because the scene is told from Elsa's point of view, Stegner has to find a way to express Bruce's terror dramatically: the child babbles incoherently, but the "visible" expression of his terror is his crossed eyes, which his older brother Chet worries will remain that way his whole life. This detail seems particularly significant—especially given Stegner's use of the "lens" image in describing the author's role in storytelling: "All you want in the finished print [that is, in the story] is the clear statement of the lens, which is yourself, on the subject that has been absorbing your attention."[5] On one level, we might imagine that Stegner cannot tell the story from Bruce's point of view because his vision of it is too "crossed," too wounded, too internalized; the finished print will be blurred. On another level, however, we might turn Stegner's metaphor around and imagine that when issues of vulnerability, control, and protection arise, Stegner's lens always crosses—that is, powerful emotions surface and entangle.

This scene concludes with a characteristic Stegner move from personal tragedy to the larger failures of life to protect "all the little live things" vulnerable to the accidents of fate, whether drowning, cancer, polio, or a host of other threats. As Elsa and the children leave Washington, they discover that the children's bunnies have fallen into a well and drowned.

A later scene on the family homestead in Montana again concerns the death of one of those little live things and implicates Bruce in his family's failures to protect. The passage is framed by references to its ostensible villain, the shrike, or butcher-bird, which kills "just for the fun of it." Bruce hates the shrike, and wants to kill it for draping a trophy sparrow on the wire fence. After a visit to some neighbors Elsa likes and Bo finds pretentious, Bruce observes his parents taunting one another; he hopes his mother will "keep still, because if she didn't she'd get [Bo] real mad and then they'd both have to tiptoe around the rest of the day" (p. 226). In this scene we are again trapped in the point of view of a character who feels a desire to protect but is powerless to do so. As the fight accelerates, Bruce tries further to intervene, interrupting his mother, "anxious to help the situation somehow" (p. 227). This time Bo decides to take out his anger elsewhere. In response to his wife's "If you shoot that harmless little bird!" and Bruce's "Don't, Pa!" he shoots a sparrow. Bruce is appalled and Elsa sarcastically remarks that "your father will want to hang it on the barbed wire" (p. 229). At the end of this scene Bruce first acknowledges that he hates his father.

This time both Bruce and Elsa have failed. Despite Bruce's best efforts to manage his parents' relationship, his father's moods, he has protected neither his mother, himself, nor the innocent bird. Indeed, the bird's death produces a feeling of loss seemingly beyond proportion for a child whose daily activities include killing various creatures. Perhaps Bruce's intense feelings of loss are connected to guilt: he recognizes that in some perverse way the sparrow has been sacrificed to protect him. He takes responsibility for its loss, asking his mother, "Ma, what'll I do with it?" (p. 228). This scene seems crucial in Stegner's evolving preservationist ethic. Throughout most of it, the boy is playing with the very gun his father uses to kill the sparrow, sighting down it, imagining killing the shrike in revenge. As he is always powerless against his father, Bruce and later Stegner characters will be powerless too against the forces of destruction his father represents. But instead of leaving trophies of death behind, they will side with the vulnerable, the wounded, will take a position of protection and responsibility. Bruce stops looking through the gun sights and starts looking through his own lens.

In one of his interviews with Richard Etulain, Stegner uses an interesting pair of images to describe the effect of his father's violence. "The effect, I'm sure, of such a dominating and hairtrigger kind of father on many kids is to breed a kind of insecurity which may never be healed. I was probably looking for security."[6] The sparrow, so secure in its life, is suddenly dead—such scenes of apparently random threat or death reverberate throughout Stegner's work. Indeed, the metaphor echoes through his work, as is suggested by a line John Daniel cites in his essay in this volume: "As surely as any pullet in the yard, I was a target and I had better respect what had me in its sights."[7] While the hairtrigger father may have been the cause for the insecurity, the mother was not able to offer the child security either. In fact, few characters in Stegner's fiction can offer anyone protection or security—"safe places"—just oases of momentary sanctuary.

In yet another example involving his mother, Stegner hints at the result of that "insecurity which may never be healed." In the letter Stegner wrote to his own long-dead mother "much too late," it is his unresolved relationship with her that leads to the wound. He describes her as "at once a lasting presence and an unhealed wound."[8] Filled with love, the essay is also filled with guilt, even shame. Among many regrets, Stegner regrets the way he told his mother's story: "I am afraid I let your selfish and violent husband, my father, steal the scene from you and push you into the background in the novels as he did in life."[9] (One is led to wonder whether Bo "stole" the two scenes I've just discussed.) Framed negatively, Stegner's wound is a gnawing guilt created by his failure to protect his mother in life and in art. Framed more positively, the wound is one of "responsibility," a value he learned from his mother. "Your morality counselled responsibility for what you did," he writes.[10] As a lasting presence, she reminds him of inevitable human failures despite our best intentions.

Here Stegner's impulse to protect leads to narrative intervention. In *The Big Rock Candy Mountain*, teen-aged Bruce sometimes leaves his dying mother alone to go out with his friends and feels guilty about it. After many years of such guilt, perhaps desiring to offer his mother the "recompense" he said he wanted to give her, Stegner revises Bruce's actions in *Recapitulation*.[11] There Bruce leaves a party to take his girlfriend Nola to meet his mother, just to please her. Yet when they arrive, he empathizes with Elsa's embarrassment at how disheveled she looks, thinking that she "deserves protection and disguise. He should have thought of that."[12] With his own scars, Bruce is quick to recognize vulnerability, to seek to protect; in retelling his story, Stegner makes the same move—and ironically confronts yet another failure.

When asked by Etulain if the life of Mary Hallock Foote, on which Stegner based much of *Angle of Repose*, reminded him of the life of Elsa Mason, Stegner responded that he had discovered "all kinds of connections" only after he finished the later novel.[13] Although he does not list his concern with the reciprocal protection between a mother and child—or grandchild—as one of these connections, surely his interest in the Foote materials had much to do with the issues of protection so central to her life, her illustrations, and her writing about being a woman artist in the West. Stegner borrows from the historical record, for instance, Foote's comment that she was "one of the protected women of her time" and her interpretation, too, of the costs of "the protected point of view" on her art and her marriage.[14] "You can't protect me from everything!" Stegner's Susan Ward cries out to her husband, Oliver.[15]

As Oliver Ward tries to protect Susan from western experience, so too does Susan attempt to protect—in some views, overprotect—her children. This theme too might have originated in the historical record, as one can see in Foote's celebrated illustrations. While "First Steps" (Figure 1) and "The Baby's Sunny Corner" (Figure 2), both set on the porch of Foote's house outside Boise and drawn to accompany poems for children in *St. Nicholas Magazine*, may seem generalized, other Foote illustrations specif-

Figure 1. Mary Hallock Foote's illustration for the poem "First Steps" (by "M.M.D."), *St. Nicholas Magazine*, April 1888, p. 449.

Figure 2. Foote's illustration "The Baby's Sunny Corner," *St. Nicholas Magazine*, May 1889, frontispiece.

ically address a mother's role in the West. "The Irrigating Ditch" (Figure 3), for instance, one of a series of ten sketches and essays called "Pictures of the Far West," published in *Century Magazine* in 1889, implies a parallel between the presence of women and their children and the promised fruitfulness of the arid West; like Stegner, Foote, the wife of an engineer, knew water was central to western life and western history. "The Coming of Winter" (Figure 4), another of the "Far West" series, suggests the precarious struggle of a homesteading family and reminds a Stegner reader of Bo Mason's attempts to feed his family through killing and freezing ducks. Perhaps the psychological depth Stegner admired in Foote's work is best suggested by "The Last Walk on the Beach" (Figure 5), from the period when Foote left her husband, who had begun to drink heavily, and lived in Victoria while she decided whether to return to her marriage—another scene that parallels Stegner's own life. The mother holds tight to her two daughters, buffeted by the wind, as a storm rolls in from the west. The sky is lovely but threatening, and the family faces a shadowy future. But Foote constructs her pyramid of figures as offering each other mutual support, even protection, against the storms of life.

Both personally and historically, Stegner had good reasons to be inter-

Figure 3. Foote's illustration "The Irrigating Ditch," from "Pictures of the Far West," *Century Magazine*, June 1889, p. 299.

Figure 4. Foote's illustration "The Coming of Winter," from "Pictures of the Far West," *Century Magazine*, December 1888, p. 163.

ested in Foote's life and to base his novel on her work. But he fictionalizes her life in order to meditate on how a storyteller, a historian, struggles with his responsibilities to his subjects. Given Foote's almost obsessive focus on the protection of children, it is intriguing that Stegner's crucial *departure* from the historical data of her life turns on a moment when a mother fails to protect her daughter—a plot change that tells us much about Stegner's own interests. As readers will remember from the narrator Lyman's speculations about the decisive moment of his beloved grandmother's life, she goes out along one of her husband's irrigating canals to meet Frank, her emotional if not her physical lover; they talk; her daughter Agnes—plucked, like the adulterous Hester Prynne's Pearl, from a rosebush—the child most like her, wanders off; and is later found floating dead in the canal. Susan is well punished in a further succession of losses: Frank, who Lyman presumes possesses the "same view of individual responsibility that Grandmother did," kills himself; Oliver pulls up the rose garden and leaves her; her son Ollie refuses to speak to her, essentially for the rest of her life.[16] Foote's life actually followed the much calmer banks of her placid "Irrigating Ditch": she never had an affair, returned to her husband, and lived

FIGURE 5. Foote's illustration "The Last Walk on the Beach," *St. Nicholas Magazine*, November 1886, frontispiece.

happily with him for another fifty years. The major tragedy of her life was indeed her daughter Agnes' death, of appendicitis, when she was a young woman.[17]

The implications of these changes—which focus on Susan's failure to protect her daughter, her loss, exposure, and guilt, are significant to our understanding of the novel, as I have explored more fully elsewhere.[18] Here they can show us how Stegner ruminates on protection through his novel's narrative technique. Like Bruce's crossed eyes, these events are only the visible signs of feelings and motives Lyman can never know. No letters exist from this period. Lyman can never enter the minds of any of the characters: "I have to make it up, or part of it. All I know is the what, and not all of that; the how and the why are all speculation."[19] Yet as Stegner chose to tell the earlier scenes from the perspective of failed protectors, so too does Lyman imagine his grandmother's point of view in the dramatic conclusion to the section:

> She never blamed her husband for abandoning her in her grief and guilt, she never questioned the harshness of his judgment, she did not turn away from those dead roses that he left her for a sign. She thought he had suffered as much as she, and she knew that for his suffering she was to blame.[20]

How does Lyman know? Why is it important to him to believe his grandmother felt this way? This passage is sharply ironic. Not so many pages earlier, struggling to assess his grandmother's life, Lyman looks to a much admired historical forefigure and says: "I shall go on writing the personal history of my grandmother, following Bancroft's advice to historians: present your subject in his own terms, judge him in yours."[21] Why does Lyman decide to "judge" Susan in her own terms?

Stegner's decision to climax Susan's life with such a "failure of protection" reverberates with his own concerns. Susan is stripped bare, exposed, yet her acceptance of responsibility dignifies her. Lyman imagines Susan Ward as feeling herself "responsible, willing to accept the blame for her actions even when her actions were, as I suppose all actions are, acts of collaboration."[22] This passage has a kind of odd Doppler effect in that Susan, who lived long before Elsa Mason, almost seems her descendant. For one of the qualities Stegner valued most in his mother was that "[her] morality counselled responsibility for what [she] did."[23] This passage simultaneously reveals Lyman's reluctance to judge his grandmother and allows him to expose her, even blame her (as indeed he has done throughout the book).

This is a characteristic narrative stance for Stegner: he exposes his characters, then offers them narrative sanctuary.

Stegner's narrative technique becomes clearer in contrast to another Foote illustration, "The Engineer's Mate" (Figure 6), and a passage from one of her novels, *The Last Assembly Ball* (1889).

> When an eastern woman goes West, she parts at one wrench with family, clan, traditions, clique, cult, and all that has hitherto enabled her to merge her outlines—the explanation, the excuse, should she need one, for her personality. Suddenly she finds herself "cut out," in the arid light of a new community, where there are no traditions and no backgrounds. Her angles are all discovered, but none of her affinities.[24]

This description of an awkward rebirth, the self "cut out" and exposed, stresses images of separation and boundary. And its linear quality reveals

Figure 6. Foote's illustration "The Engineer's Mate," for "The Conquest of Arid America" by Willam E. Smythe, *Century Magazine,* May 1895, p. 90.

the illustrator's eye: her outlines no longer "merged" with those around her, her personality no longer "explained" by her background, she becomes a "cut-out" figure whose three-dimensional qualities, her depth, her "affinities," remain undiscovered.

Although published separately, "The Engineer's Mate" might have been drawn to illustrate this passage about isolation and exposure. The drawing contrasts the Eastern "tradition," with all its cultural baggage and fancy clothing, with Western space, vast and empty, though defined by the engineer's telegraph wires and railroad tracks. Preposterously out of place, the "cut-out" figure of the woman is framed by her husband's occupation, the perspective established by his "lines" pulling her into the far distant emptiness, her identity labeled by his career. (Her own career is in evidence, however, in the fold-up easel that is part of her baggage.) This self-portrait is simultaneously humorous and threatening, revealing Foote's ironic sense of her own absurdity and an anxiety about being observed, vulnerable, and alone. She offers herself no protection.

Instead of presenting Susan Ward in this way, cut out, exposed, Stegner offers her a narrative frame, a character who vacillates between being a judgment man and a mercy man. From exploring Susan's guilt, from her own point of view, Lyman moves to separate himself from those who have judged her:

> Nevertheless, I, who looked up to [my grandfather] all his life as the fairest of men, have difficulty justifying that bleak and wordless break, and that ripping-up of the rose garden, that was vindictive and pitiless. I wish he had not done that. I think he never got over being ashamed, and never found the words to say so.[25]

In Lyman's projections, Oliver's shame results from a pitiless failure to forgive and accept. If Susan cannot defend herself, Lyman will question the harshness of the judgments he imagines leveled against her, partially by accepting blame himself. Yet given his tragic reading of his grandmother's life, surely this is yet another example of failed protection.

As much as Stegner—his implied author, his narrators—would like to protect others, he recognizes the danger. "Pitying others indiscriminately," Stegner writes, "we are pitying ourselves, and there is no more romantic and dangerous kind of moral obfuscation than pity, or self-pity, gone out of hand."[26] He almost quotes himself near the end of *Crossing to Safety*.

After a long discussion with some of the Lang children about how he should write a book about their lives, perhaps recalling his desire to protect their father, Larry Morgan asks, "I wonder if I could recreate any of us without my portraits being tainted by pity or self-pity."[27] *Crossing to Safety* explores how to achieve a balance in judging other's lives.

The novel's beautiful but baffling title, which comes from a poem by Robert Frost called "I Could Give All to Time," addresses this struggle. The poem's narrator questions:

> But why declare
> The Things forbidden that while the Customs slept
> I have crossed to Safety with? For I am There
> And what I would not part with I have kept.

This speaker does find sanctuary "there." But he is unable or unwilling to "declare" what he carried with him, what he "would not part with." Larry, and his creator, carry with them that ethic of protection: the understanding that all we can offer one another is solace for our inevitable wounds.

The protective impulse, once again on multiple levels, is at the heart of *Crossing to Safety*. Charity Lang reveals the dark side of protective impulses; she uses the desire to protect to disguise her willful need to control others' lives. In "protecting" her husband from the pain of her loss, the inevitability of death, she denies him the right to his own need to offer her a protective love, failed though it may be. Nor can Larry protect his friend Sid from his complicity, his "acts of collaboration," in his marital failures, from being the man he is. There is no sanctuary even in Eden. Unable to protect his wife from polio, Larry Morgan spends the rest of his life trying to achieve a balance between caring for her, yet another wounded creature, and respecting her need for independence. (One thinks here, too, of Joe Allston's desire to protect Marion in *All the Little Live Things*.)

Larry struggles to achieve that same balance in his narrative stance. Like his creator, Larry Morgan might well believe that he would not protect his friend in the story he tells; he admits, for instance, to feeling ashamed for Sid. But he also acknowledges pitying him, and Stegner's description of writing his novel tells us how the narrative struggle to understand can become intertwined with the impulse to protect:

> I suppose I wanted to justify their lives, bring them together, lay their ghosts.
>
> In that effort I wrote very close to memory and fact. I resisted

> whenever I felt myself wanting to adjust or improve or straighten out. . . . What I wrote was a labor of love and bafflement.[28]

Larry writes as a sympathetic friend. He achieves a balance between pity and honesty, between care and the recognition that ultimately we cannot offer our friends—or characters—"safe places."

STEGNER'S PROTECTIVE IMPULSE extends into his nonfiction writing as well. Much of his writing about western literature and history can be read as an effort to protect the West against the mythologizers. His books on the Mormons are historically honest, sometimes critical, but written partly to defend them against stereotyping as either Saints or Zealots. His attitude is best captured in an image from *Recapitulation*, when Bruce Mason revisits Salt Lake City and his memories: "Though [Bruce] found that he couldn't admire [the Mormon Temple] architecturally, it struck him as comforting and safe—he felt protective about it."[29] And most significantly, Stegner finds the impulse to protect at the heart of the American conservation movement. The most often repeated words in "A Capsule History of Conservation" are: protection, preservation, sanctuary, reserves. He honors "preservationists" like Muir, says "wilderness preservation" is one of the best ideas America ever had. Yet the constellation of ideas that he sees as orbiting protection ultimately influence the way he argues for preservation. In arguing for the need for wilderness areas in "Wilderness Letter," for instance, he presents wilderness as a spiritual rather than a literal sanctuary: "The reminder and the reassurance that it is still there is good for our spiritual health, even if we never once in ten years set foot in it."[30] No doubt influenced by Leopold, he also argues for the preservation not just of pristine wildernesses but of "wounded" areas: "in a dry country such as the American West the wounds men make in the earth do not quickly heal. Still, they are only wounds; they aren't absolutely mortal. Better a wounded wilderness than none at all."[31] Balancing pragmatism with idealism, he knows that protection will inevitably fail, that wounds will be inflicted, but our spiritual health, he says, lies in the idea of preservation.

TO CLOSE, I offer three quotes to ponder. The first comes from a book Stegner much admired. Terry Tempest Williams' *Refuge* is about a threatened bird sanctuary, which recalls Bruce Mason's desire to offer refuge to a sparrow. One scene in particular seems to speak elliptically about Stegner's work. Out walking with her mother and grandmother, Williams sees a killdeer fluttering around. It is apparently wounded, but Williams offers this interpretation: "We must be close to its nest. She's trying to distract us. It's a protective device."[32] As the pitiful "trophy" sparrow of *Big Rock Candy Mountain* is a symbol of cruelty and destruction, this killdeer is symbolic of efforts to preserve life.

Perhaps the killdeer offers a metaphor for a narrative stance. Describing the role of the storyteller, Stegner writes:

> Some writers want to expose themselves, some to disguise themselves, some to efface themselves. Some who appear to expose themselves are distorting themselves for reasons of their own. There is more than one way to impose order on your personal chaos; but since good writers write what is important to them, they are bound to be in there somewhere, as participants, or observers or ombudsmen.[33]

Or, I would add, as protectors.

Stegner's friend Robert Frost called the imposition of order on personal chaos a "momentary stay against confusion." Writing from the protective impulse, Stegner offers a momentary stay against modernist despair. Through the protective Joe Allston, yet another bird-watcher, the protagonist of Stegner's story "A Field Guide to Western Birds," and the narrator of *The Spectator Bird*, Stegner provides a hopeful vision of refuge and finally an example of reciprocal solace for life's inescapable wounds:

> It is something—it can be everything—to have found a fellow bird with whom you can sit among the rafters while the drinking and boasting and reciting and fighting go on below; a fellow bird whom you can look after and find bugs and seeds for; one who will patch your bruises and straighten your ruffled feathers and mourn over your hurts when you accidentally fly into something you can't handle.[34]

CHAPTER FIVE

WALLACE STEGNER'S PRACTICE OF THE WILD

JAMES R. HEPWORTH

I

ALTHOUGH THE BEST of his critics, such as Jackson Benson, freely acknowledge that Wallace Stegner and his writing were all "of one piece," it is too little appreciated, I think, how Stegner's ideas about conservation and ecology operate in his fiction. Even Stegner himself expressed doubt on the subject. "I have been unable to bring much of my thought about conservation into fiction because I suspect myself when I begin to be doctrinaire," he once told me. "I guess I must hold the integrity of the material to be of greater value than any message that I might want to get across. If the material itself dictates that message it would be in there, but I don't seem to be able to put it in by force or will because that seems to me a dilution of the essential."[1]

But far from complicating matters, as this statement might first appear to do, Stegner's words reveal what must eventually become self-evident to any creative writer: "The conscious agenda-planning ego occupies a very tiny territory, a little cubicle somewhere near the gate, keeping track of some of what goes in and out (and sometimes making expansionist plots), and the rest takes care of itself. The body is, so to speak, in the mind. They are both wild."[2] Our consciousness is the world that surrounds us. In other words, as Stegner fully realized, the conscious recesses and subconscious deeps of our minds are wilderness areas. To know as much, Stegner argues, we need only witness any one of the great reaches in our few remaining wildernesses. Here he cites the Robber's Roost country in Wayne County, Utah, before the Four Corners power plants polluted the view. Notice the connections he makes not only between inner and outer space but between body and soul, and notice, too, that his reference to "wilderness" automatically invokes the sacred.

> It is a lovely and terrible wilderness, such a wilderness as Christ and the prophets went out into; harshly and beautifully colored, broken and worn until its bones are exposed, its great sky without a smudge or taint from Technocracy, and in hidden corners and pockets under its cliffs the sudden poetry of springs. Save a piece of country like that intact, and it does not matter in the slightest that only a few people every year will go into it. That is precisely its value. Roads would be a desecration, crowds would ruin it. But those who haven't the strength or youth to go into it and live can simply sit and look. They can look two hundred miles, clear into Colorado; and looking down over the cliffs and canyons of the San Rafael Swell and the Robber's Roost they can also look as deeply into themselves as anywhere I know.[3]

And finally, notice that the emphasis in the passage is not upon division, a separation of worlds ("Technocracy" and "wilderness"), but rather upon unity and continuity, the similarity between human and nonhuman, a distinction the closing lines obliterate. In *Angle of Repose*, a wheelchair-bound Lyman Ward (who has neither the youth nor the strength to "go into it") tells us, "As I look down my nose to where my left leg bends and my right leg stops, I realize that it isn't backward that I want to go but downward. I want to touch once more the ground that I have been maimed away from."[4] As Elliott West points out in his essay in this volume, Lyman is referring to the remembering of himself and the earth as well as to remembering (historiography) as a duty. These internal and external wilderness areas where people can look "deeply into themselves" are the primary settings of Stegner's most successful novels, all of which marry their inner to their outer worlds and vice versa. Here again, however, our consciousness is the world that surrounds us. We can see this conviction manifest itself in the first line of Stegner's last novel, *Crossing to Safety*: "Floating upward through a confusion of dreams and memory, curving like a trout through the rings of previous risings, I surface. My eyes open. I am awake."[5]

If you really want to see how Wallace Stegner's ideas about conservation operate in his fiction, you must do what Larry Morgan does in *Crossing to Safety*: wake up and enter the wilderness of human thought. You have to become aware. You have to become vulnerable. You have to offer yourself up to the experience of discovery and any of the disagreeable truths that Stegner may reveal. In short, you have to be willing to confront what Stegner's Joe Allston in *All the Little Live Things* calls "existential problems" and "ultimate questions": "Who am I? How to be? What is the meaning of everything?"[6] And the first question I would have you confront is this: do you genuinely believe, as Wallace Stegner did, that you are an animal?

The idea that human beings are animals is at last being introduced in our nation's schools. Yet it is an idea most of us have been spiraling around, testing and investigating in circular fashion, all our lives. If, like Wallace Stegner, we were lucky enough to grow up as a "sensuous little savage" in urban and rural places with, say, the snowy front range of the Wasatch Mountains constantly in view, then we have also had the good fortune to witness humans and other animals continually interacting within the same realm. But some of us—maybe even most of us—still feel remote from the nonhuman world. We are still unsure we are animals. We want to believe we are something new, something different, something better. But what? For lack of a better term, let's just say "an American." But an American, according to Wallace Stegner, "insofar as he is new and different at all," is a civilized human being who has "renewed" himself "in the wild."[7] Now think of that trek through the Vermont outback the two couples make in *Crossing to Safety*, or of wheelchair-bound Lyman Ward and his arduous daily race on crutches past the pines of Zodiac Cottage, or Bruce Mason's shotgun marches after blue-winged teal in *The Big Rock Candy Mountain*. But think, too, of the way Stegner structures his fiction and recall the wild characters who populate our minds in the act of reading it.

Wild characters? Bourgeois, middle-class, anachronistic, old-fashioned Wallace Stegner? Yes, indeed. Despite their civility, Stegner's narrators tend to be revolutionaries who have history and civilization on their side. Their thoughts and their actions "fly in the face of accepted opinion and approved fashion" by reasserting "values so commonly forgotten or repudiated that, re-asserted, they have the force of novelty."[8] Indeed, "by every stereotypical rule of the 20th century," both Stegner's narrators and his novels should be dull because they are so quiet. Certainly the image of the artist in the twentieth century is still largely the legacy of romantics who misread their Milton. Unlike the romantics, Stegner did not believe Milton was of the devil's party without knowing it or that an artist has to be rabid—"crazy or alcoholic or suicidal or manic-depressive"—to be "a legitimate spokesman to the world."[9] Stegner's narrators are in fact generally wild animals in an altogether different sense. They are creaturely social beings intimately in touch with their surroundings, as observant as deer, as familial as wolves. Our American idea of wilderness and what constitutes the wild, however, has yet to enlarge itself enough to encompass Wallace Stegner's vision of them—or even the truth about ourselves. We still tend to view him and his work as outdated and provincial. At best we see him as a regional writer, a "transitional figure," or a misplaced nineteenth-century "realist." At worst, we see him as a sexist, a racist, and a plagiarist.[10] But in fact Stegner is a postmodern storyteller with postmodern concerns. His contemporary West is not a marginal region but the mainstream, "the United States, only more so."[11] In other words, Wallace Stegner's vision is inclusive. Like the geographical settings of his novels, it is literally continental and international in scope.

Although it has yet to be properly documented, one of Stegner's greatest contributions to twentieth-century letters is his reassertion of the Native American paradigm of nature as familial and relative rather than alien or foreign. One of the most common themes of his fiction is marriage and family. But not merely marriage between people: marriage, rather, between people and their places and their communities. And certainly not just between one human community and another but marriage between human communities and nonhuman communities, both plant and animal. To be sure, "that romantic atavist we sometimes dream of being, who lives alone in a western or arctic wilderness," does find its objective correlatives in Stegner's characters. We recognize aspects of the atavist in Stegner's fictional isolates like Lyman Ward and the Bruce Mason of *Recapitulation*. This romantic atavist perhaps incarnates himself most completely in figures like Bo Mason or that unruly young man, Jim Peck, who constructs the tree house in *All the Little Live Things*. But as Stegner points out, this "wild man of the woods" has "many relatives who are organized as families."[12] Like Bo and Elsa Mason in *The Big Rock Candy Mountain* or Susan and Oliver Ward in *Angle of Repose* or Ruth and Joe Allston in *The Spec-*

tator Bird or Sally and Larry Morgan in *Crossing to Safety*, these families "drag their exposed roots with them" across the North American continent and sometimes all the way to Mexico, Canada, and Europe.

Although the settings of Stegner's fictions are often in fact "western," we ought to keep in mind that *Crossing to Safety* never moves west of Madison, Wisconsin, much less beyond the 100th meridian. And—who knows—perhaps to emphasize the point that he lived about a third of his life as a member of those "academic tribes who every June leave Cambridge or New Haven for summer places in Vermont, and every September return to their winter range,"[13] Vermont is the place where Wallace Stegner's ashes now mingle with the compost of beech trees. At least half of *The Spectator Bird* is set, not in California, but in Denmark where that wild African Dane, Karen Blixen, makes two appearances as sudden as one of the Weird Sisters out of *Macbeth*. The primary geographical setting for both *Angle of Repose* and *All the Little Live Things* is California at the height of the cultural revolutions circa 1970. But much of *Angle of Repose* traverses the continent from Grass Valley to New York, from New York to New Almaden and Leadville and Boise, and from Boise across the Mexican border to Michoacán. . . . And so it goes, too, with the novellas and short stories—for anyone who cares to study them.

Like aboriginal Americans, Stegner's protagonists tend to follow "local custom, style, and etiquette without concern for the standards of the nearest metropolis or the nearest trading post."[14] And I don't mean just the self-reliant, independent cowboys or the wolfers in Stegner's early fictions like "Genesis," or the freedom-loving, self-reliant roving males like Bo Mason. I'm talking about something else. I'm talking about the insubordinate, subversive, unruly, and occasionally unrestrained behavior of Stegner's seemingly most domestic citizens like Bruce Mason, Lyman Ward, Susan Burling Ward, Charity Lang, and Joe Allston—stubbornly self-reliant, proud, and independent cusses who fiercely resist oppression, confinement, or exploitation, whose thoughts, speech, and actions Stegner crafted to appear artless, spontaneous, wild, and free, people who are privately highly expressive and openly physical and even sexual but publicly stoic and reserved.

The first one who comes to mind, of course, is fifty-eight-year-old Lyman Ward, who, during one scene in *Angle of Repose*, must cover his lap with a blanket in the presence of his earthy and intelligent Girl Friday, Shelly Rassmussen, in order to conceal his erection. "Everything potent,"

Stegner writes, "from human love to atomic energy, is dangerous; it produces ill about as readily as good; it becomes good only through the control, the discipline, the wisdom with which we use it." In the same paragraph he goes on to write, "Much of this control is social, a thing which laws and institutions and uniforms enforce, but much of it must be personal, and I do not see how we can evade the obligation to take full responsibility for what we individually do. Our reward for self-control and the acceptance of private responsibility is not usually money or power. Self-respect and the respect of others is quite enough."[15] In dramatizing Lyman Ward's control over his erotic attraction to Shelly, Stegner reminds us of our own private responsibilities, but he also reminds us that our bodies are wild. Their behavior, as menstrual cycles, erections, nightdreams, and spontaneous abortions demonstrate, is sometimes far beyond our control. Human conduct—how we decide to act and think—is another matter.

In his 1964 essay for *Saturday Review*, "Quiet Crisis or Lost Cause," Stegner refers to "the real conservation problem"—overpopulation, a direct result of our misconduct of ourselves.[16] In 1930, the world's population was 2 billion people. By 1975, it had doubled to 4 billion. Current projections put the figure in excess of 10 billion by the year 2030, an increase of 500 percent in just one century. But in 1964 Stegner noted that the projected world population for 2150 is 150 billion people. As his biographer observes, that will mean "fifty times as many people and problems" as we have now.[17] "It will no longer be human life," Stegner warns us, "it will be a termite life." Even with proper conduct, Stegner argues, "you cannot turn well over a hundred million annual visitors into a park system and expect them to leave no marks, even if they are well-intentioned, well-educated, and well-policed." As for the national parks and the nation's wilderness areas themselves, he views them as essential to the survival of human life:

> They are the indispensable opposite to industrial regimentation and dehumanization; in our haste to become something new, termites or otherwise, we need constantly to be reminded of who we are: creatures, made of water and chemicals but the children of sun and grass, and cousin by warm blood to birds and mammals. Though it is necessary for our survival to husband resources, it is necessary for our emotional health to husband natural things, and places where they may be known.[18]

Twelve years after these wrote the words, Stegner published *The Spectator Bird*, which features that attractive, jocular, tennis-playing Dane,

Count Eigil Rodding, in his "corduroy jacket and jodhpurs and an Ascot tie."[19] He and his father before him, both world-class genetic engineers, have long known how to grow crops and animals "by blueprint."[20] In his "Wilderness Letter," Stegner writes, "We are a wild species, as Darwin pointed out. Nobody ever tamed or domesticated or scientifically bred us."[21] But Count Eigil Rodding sees no reason why he, fully armed Cultural Hero, should not play Dr. Faustus and do just that. To manipulate the gene pool for size, strength, speed, intelligence, good digestion, and virility, he has copulated with his half sister and their daughter. Now in his mid-forties, Eigil looks forward with his "amber eyes" to his golden years when he can rape his granddaughters.

In this sense, then, *The Spectator Bird* is an environmental novel. It is a book that adds new meanings to fashionable terms like "ecoterrorism" and "deep ecology," a novel that gives a few little twists of its own to those worn-out phrases "crimes against nature" and "population control." One of the crises the novel's protagonist, Joe Allston, survives is his erotic attraction to Eigil Rodding's sister, the Countess Astrid. To be sure, it is an emotional crisis, a crisis of marriage and family, but a crisis nevertheless intimately related to the "quiet crisis" of overpopulation. And human conduct is very much at issue here. Eigil is a certified genetic engineer, one who would "sanitize" the world by taking every advantage away from nature. He wants to scientifically breed human beings to become pure, tame, and domestic, the same way he does his corn and his cattle. Like H. G. Wells' Dr. Moreau, Eigil has no qualms when it comes to playing God. "Evil," Joe Allston, that Spectator Bird, confides to his readers, "if it exists, is not all lumpy and ugly like a toad. It is often more attractive than what people call good."[22] Like the count himself, the idea of manipulating the human gene pool is "attractive." Doing so could "improve" the species, extend human life expectancies, and perhaps alter human behavior so that we could do away forever with that troublesome problem of free will.

But when I say that Stegner's protagonists tend, like aboriginals, to "follow local custom, style, and etiquette without concern for the standards of the nearest metropolis or the nearest trading post," I mean that they follow what Gary Snyder calls "the etiquette of the wild world"—a world that "requires not only generosity but a good-humored toughness that cheerfully tolerates discomfort, an appreciation of everyone's fragility, and a certain modesty," a world that calls for "self-abnegation" and "intuition."[23] Count Eigil Rodding wants to rid his fiefdom of a wild stag because the animal has a deformed set of antlers and has managed to elude and outsmart him. Consequently, he seeks out the adventure of killing the animal with a rifle and drags Joe Allston along with him. But like the people of indigenous cultures Snyder describes, Stegner's protagonists "rarely seek out adventures." When they do deliberately risk themselves, it is "for spiritual rather than economic reasons."[24] Thus, in the wake of grief for their son, Joe and Ruth Allston sail to Denmark and, years later, risk reading aloud Joe's Denmark journal with its hint of marital infidelity. Lyman Ward, in order to examine what happened to his adventurous grandparents, Susan and Oliver, moves into his ancestral home where he mines the past in order to understand the present. These people have the same quiet dignity that characterizes many Native American elders. Snyder quotes from an interview between a young woman anthropologist and a contemporary Haida elder: "What can I do for self-respect?" the Haida elder asked, repeating the anthropologist's question, and then answered: "Dress up and stay home."[25]

The search for an ancestral home runs like a bright orange thread through the entire warp of Stegner's fiction, although the "home" is, as Snyder observes, "as large as you make it."[26] That search is what sends the world-weary ambassador, Bruce Mason, back to Salt Lake in *Recapitulation* just as surely as it sends Peck's bad boy, Jim, in *All the Little Live Things* literally up into the evolutionary treetops from which our wild ancestors theoretically descended more than a million years ago. Stegner's characters are more migratory than geese. On the subject of home, Stegner is fond of quoting his friend Robert Frost: "Home is where, when you have to go there, they have to take you in." But more often in Stegner's novels home is what you can carry with you. And what we carry with us, naturally, is the seed of something primitive and wild: civilization—which is to say, our culture, the total product of our human creativity and intellect.

Anyone who thinks Stegner treats the idea of wildness as an attribute confined to those portions of the West that are relatively free of human influence and history should read deeper in the Stegner canon. Far from limiting his idea of the wild to that 2 percent of formally designated wildlands in the United States, Stegner insists that wildness is everywhere. "Going to the kitchen to make coffee," Joe Allston discovers "en route that the clerestory windows above the bookcase wall in the living room" are "leaking" and spends "half an hour on the stepladder taking down kachina dolls, papier-mâché Hindu gods, Hopi bowls, and other bric-a-brac from the drowned top shelf, sponging up a bowlful of water mixed with the cobwebs, dust, and dead flies."[27] Joe makes "one pass across the rug" with the vacuum cleaner "and pop, the cleaner's howl" dies and all the lights go out.[28] A single supercharged meteorological burst from the wild atmosphere displaces Joe and Ruth in their own home. Joe recalls that one day "last winter the power was off nearly all day," so long that he and Ruth "paid three different calls to people" they "didn't especially want to see, just to get to use a bathroom."[29]

Like the migrant but nevertheless placed people they are, Joe and Ruth Allston cheerfully tolerate the discomfort of their temporarily nonelectric house. Joe, who is something of a "preparedness freak," appears even to welcome the inconvenience and goes off down the hill to unplug the culvert in the downpour while Ruth prepares to make corn fingers, apricot soufflé, and chicken breasts amandine on a "little tin stove" lit by two cans of Sterno. These she will serve to the visiting Italian novelist, Césare, whom they have invited out to Chez Allston *en famille*, following local custom and etiquette without concern for Césare's cosmopolitan European standards. Césare would prefer to dine at a restaurant in the nearest metropolis, whereas Joe and Ruth prefer to keep their self-respect by dressing up and staying home. True: the temple of the Allston home isn't exactly the primary wilderness temple in *Crossing to Safety* where Larry and Sally Morgan and Sid and Charity Lang entertain each other in the Vermont outback by eating undercooked chicken, sleeping on rocks, and skinny-dipping at sunrise, but neither is it the Dutch Renaissance castle in Denmark where the countess' "wicked brother," Eigil, conducts his incestuous genetic experiments. "Every region has its wilderness," Gary Snyder contends, "there is the fire in the kitchen, and there is the place less traveled."[30]

Joe and Ruth Allston are placed people who follow local customs. The stories Joe tells, like the stories of Stegner's other first-person narrators, build what anthropologist Keith Basso in referring to the Western Apache narrators calls "place-worlds."[31] The story Joe Allston tells in *All the Little Live Things*, for instance, is mostly a story about what happened at the places he names: "the cottage" where Marian Catlin lived with John and began to raise their daughter before she died of cancer; the LoPresti place; the Weld place; "the bottoms" full of poison oak where Jim Peck built his treehouse and his ashram and his dump. Joe and Ruth are attempting to resolve their culture's dichotomy between the civilized and the wild—to become full human beings by embracing rather than rejecting the organic world. The simple act of walking is their great adventure, their meditation, the means they use to balance spirit and humility, an act, not unlike Joe's gardening, that offers them extraordinary teachings "of specific plants and animals and their uses."[32] These teachings often run counter to the values of the counterculture as well as the Allstons themselves. Think, for instance, of that scene in *All the Little Live Things* where Joe shatters the silence in his garden with "an appalling blast" from his shotgun so that the "two-foot circle of earth" on which he has been concentrating "splashes like water." Then he uproots something out of Beowulf: "a fat old gopher with a big head and naked-pale feet: *Thomomys bottae*, the Evil One."[33] Later, however, Joe mistakenly exterminates a king snake, only to discover that the snake has entered the garden with no other purpose in mind than to eat gophers.

In *The Spectator Bird*, the Allstons' maid, Minnie Garcia, gleefully reports having seen what in my subdivided part of the world has become a common sight nearly every spring. On her way to work, Minnie drives "over the hill past one of the new tracts just in time to see one of the bulldozed shelves let go its hold and slide smoothly down into the creek, leaving the aghast residents staring from the rain-swept edge of what had once been their front yard . . . fence, trees, part of the lawn, the whole business."[34] In that same novel, too, we can't help noticing that the Allston Eden, which has become "too tamed" for mountain lions, has "an ungulate problem worse than Yellowstone's." The deer "come in, twenty at a time," Joe reports, "to sleep in our shrubbery and eat our pyracantha berries, roses, tomatoes, crabapples, whatever is in season."[35] But one of the conclusions we must draw from Joe's descriptions of a world where houses hang from the hills is that we have overpopulated ourselves. We have yet to become "good animals" capable of living fully in the places we love without destroying them—and ourselves.

In Stegner's view, as in Gary Snyder's, the body of the nation is, so to speak, in the mind of the nation, and they are both wild. That house Minnie describes teetering on the brink of extinction is a ghost house, a fiction, but its meaning is wild and subject to multiple interpretations. For me it represents the statehouse of a nation attempting to own it all and control it all and conquer it all; a nation willing, like the Vietnam commander, to destroy the village in order to save it, not only refusing to set aside and preserve those sacred places where men and women have always been no more than visitors, but refusing as well to recognize, preserve, and maintain the local and regional commons. Each day we move closer to Count Eigil Rodding's paradigm.

"You know what I miss on this marvelous estate of yours?" Joe Allston asks Eigil Rodding in *The Spectator Bird.* "What?" Eigil returns, grinning. "Wild things," says Joe. "Little cottontails or gophers or snakes or moles or raccoons or polecats that breed in the hedges and live in spite of you. Holsteins and short-haired pointers are nice, but a little predictable."[36] Aldo Leopold, in his contemplation on "the hidden meaning in the howl of the wolf," warned against such predictability in his essay "Thinking Like a Mountain": "We all strive for safety, prosperity, comfort, long life, and dullness. . . . A measure of success in this is all well enough, and perhaps is a requisite to objective thinking, but too much safety seems to yield only danger in the long run. Perhaps this is behind Thoreau's dictum: In wilderness is the preservation of the world."[37]

All his life Wallace Stegner held onto the idea of Creation as something sacred. More than once he laughed aloud at my youthful attempts to "improve" the world. I don't believe he thought it could be done. Instead, I believe he held with Lao-tzu:

> The world is sacred.
> It can't be improved.
> If you tamper with it, you'll ruin it.
> If you treat it like an object, you'll lose it.[38]

Human beings, by contrast, can improve and enlarge themselves. They can work toward the perfectibility of their personal and collective social order. All of us are (or ought to be) in what Stegner humorously called "the self-improvement business" where "environmentalism or conservation or preservation, or whatever it should be called, is not a fact, and never has been. It is a job."[39] For Stegner, Native American attitudes toward the earth are "healthier than white attitudes."[40]

And that is what Stegner's protagonists are mainly after: healthy attitudes, self-knowledge, self-respect, and self-improvement—an enlarged understanding of their proper place in a universe where, if God exists, then, to use Wendell Berry's phrase, "He is the wildest being in Creation."[41] Stegner's narrators and characters attempt to come to terms with the realities of disease and death, to establish ways of living collectively and communally with other wild communities. Inescapably, many of them challenge the ideas of their creator by trying to dominate nature rather than coexisting with it. That good man, Oliver Ward, the dam-building civil engineer in *Angle of Repose,* is a case in point. The shotgun-wielding Joe Allston is another. But Allston learns from his mistakes because "he conducts his literary explorations inward, toward the core of what supports him physically and spiritually." Like his creator, Joe is one of the "lovers of the known earth, known weathers, and known neighbors both human and nonhuman" and so belongs to that honorable tradition of "placed" persons that includes great names like Crazy Horse, Sitting Bull, Thoreau, Cather, Frost, and Faulkner.[42] Like Lyman Ward and Sally Morgan and Susan Burling and Bruce Mason, Joe knows his own history and learns from it because he has searched it out and studied it instead of throwing his history overboard into the Atlantic or the Pacific, as so many other Americans have done. If Joe Allston becomes an "immortal," like Isak Dinesen's Donna Elvira, it will be because Wallace Stegner shaped him out of reality to be "vivid and true to fact and observation."[43]

But at least part of Joe's meaning, if we want to call it that, derives not from his luminous character but from the structure of Stegner's story. If we want to participate fully in the story's meaning, we have to imagine and help rebuild the same place-worlds Stegner constructs. Still, we might want to observe that the structure of Stegner's story is all process—all steps and missed steps, actions and reactions—but operations that nevertheless produce natural changes that permit Joe to pass from one set of emotional conditions into others. I'll come back to the subject of structure in Stegner's fiction in a moment. But first I want to point out that gradually, through the course of two novels, Joe moves, slowly and cautiously, to be sure, from states of doubt—from states of physical, moral, and spiritual uncertainty about himself and the world—into states of enlarged understanding. By the time we reach the end of *The Spectator Bird*, Joe is no longer just one of those to whom life happens: a powerless victim of old age with its plagues and humiliations. On the contrary, he is a full and active and hopeful participant in the rites and rituals that celebrate life's mysteries. At the end of the second novel Joe walks with his wife out into the darkness in hopes of catching sight of a rare lunar rainbow. They don't, of course. But that doesn't matter. What matters is that by unsystematic trial and error Joe has negotiated the wilderness of his emotions to recreate himself as one example of that New Man, the American, committed to his physical and social surroundings, realistically hopeful, linked to his ancestral life, which continues within him, but also linked to contemporary society and all the little live things, the wild creatures missing on Eigil's systematic and genetically engineered Danish island. Count Eigil Rodding, on the other hand, presents us with a model that violently rejects the organic world and substitutes for the reproductive universe the paradigm of a sterile mechanism and a production economy. Under the guise of making the planet safe for human beings, Eigil wants to take the advantage away from nature, to control and even eliminate the wild. Like any ruler-scientist-engineer, he tinkers with the powers of death and life to control nature and thereby degrades the entire earth.

In Stegner's terms, Joe Allston has become "a good animal"—part of the environment of trees and rocks and soil, brother to the other animals, part of the natural world and competent to belong in it."[44] The human family is a society. But as anyone who has ever studied chickens, wolves, coyotes, whales, bees, and communities of plants will tell you, the social order pervades the natural world. Like the earth, the social order itself is wild and sacred. At the novel's conclusion, Joe Allston recognizes as much:

> The truest vision of life I know is that bird in the Venerable Bede that flutters from the dark into a lighted hall, and after a while flutters out again into the dark. But Ruth is right. It is something—it can be everything—to have found a fellow bird with whom you can sit among the rafters while the drinking and boasting and reciting and fighting go on below; a fellow bird whom you can look after and find bugs and seeds for; one who will patch your bruises and straighten your ruffled feathers and mourn over your hurts when you accidentally fly into something you can't handle.[45]

This extended metaphor marries darkness and light, the inner and the outer life, male and female, accident and design, freedom and civility. It makes no distinction between the wild and the domestic. Indeed, in these few lines we witness Wallace Stegner's insistence upon the wild and our need to recognize ourselves as wild social creatures. Here, too, we can see in his metaphor of marriage Stegner's insistence upon the Native American paradigm of nature as familial, relative, and communal. It is a paradigm that recognizes what Gary Snyder calls "the play of the real world, with all its suffering, not in simple terms of 'nature red in tooth and claw' but through the celebration of the gift exchange quality of our give and take," a paradigm that quietly acknowledges "that each of us at the table will eventually be part of the meal."[46]

II

According to the novelist John Gardner, "the artist's primary unit of thought—his primary conscious or unconscious basis for selecting and organizing the details of his work—is genre."[47] If this is true, then perhaps the way a writer works through various genres can also tell us much about what he is thinking. In other words, form often turns out to be content. The medium is at least part of the message. To phrase the matter still another way, by employing the form of the journal in his novels, Stegner plays with the nature, origins, and process of American literature, as well as the nature of American life. To fully explore this idea in relationship to Stegner's fiction, of course, would require another essay. Here I simply want to suggest one more way that Stegner employs the idea of wilderness in his last novels, the majority of which, as we have seen, are set in the minds of their narrators who often build their place-worlds in the very places where their stories happen.

"What is added to brute fact by art," Stegner reminds us, "is something like what is added to the bumblebee to permit him to fly. Aerodynamically it is impossible that a bumblebee should fly; in experience, he buzzes by your ear."[48] So far as I know, no critic has paid much attention to Stegner's mechanics or aerodynamics—to his architecture, the forms and structures of his novels, to the way he fluidly crosses literary boundaries and genres. Stegner's nonfiction novel *The Preacher and the Slave* (1950) predates Truman Capote's *In Cold Blood* (1965) by fifteen years and the "New Journalism" of Tom Wolfe by nearly two decades.[49] But it is Wolfe and Capote and Norman Mailer who get the credit for originating a "new" American genre that was already fully formed in Dreiser and Twain. No critic has yet paid any attention to the way Stegner hybridizes his own forms from the genetic materials of the Bildungsroman, the Dime Novel, the Historical Novel, the Kunstlerroman, the Psychological Novel, the Anti-Novel, the Antirealistic Novel, the Picaresque Novel, the Science Fiction Novel, and the Roman À Clef. In *Angle of Repose* Stegner was doing something no other American writer was doing at the time: employing all the techniques of the "chaos-drunk" writers of metafiction against themselves to produce a highly original but also highly traditional classic.

In theory, at least, the journal form that Stegner repeatedly employs in his California novels and parts of *The Big Rock Candy Mountain* has one great and distinct advantage over all other literary forms. Put in simple terms, like the American "anthology" Stegner describes in his essays, the journal form is inclusive rather than exclusive. Consequently, it can accommodate diverse styles and points of view and in fact encourages the writer to employ them. But first consider briefly what Stegner has to say about the diversity of American life:

> Though we are swept by fads, every fad breeds its counterfad. Our lifestyles are so ferociously various that our reaction to a style not our own—and we may find it next door or in our own family—may be anything from apathy to disgust. It is not simply a polarization between young and old, though that is part of it and always has been. There are not two sides to every question, but a dozen or a hundred. American life is not a conflict between conservative and liberal, black and white, establishment and counter-culture, work ethic and pleasure principle. It is a blindfold battle royal. You may hit anybody you run into.[50]

As Stegner describes it, American life is a kind of homemade world (to borrow Hugh Kenner's phrase) without clearly defined boundaries and strict conventions. At issue in Stegner's definition is the notorious American idea of "freedom." Now compare Tristine Rainer's description of journal writing:

> There are no mistakes. You cannot do it wrong. At any time you can change your point of view, your style, your book, the pen you write with, the direction you write on the pages, the language in which you write, the subjects you include, or the audience you write to. You can misspell, write ungrammatically, enter incorrect dates, exaggerate, curse, pray, brag, write poetically, eloquently, angrily, lovingly. You can paste in photographs, newspaper clippings, cancelled checks, letters, quotes, drawings, doodles, dried flowers, business cards, or labels. You can write on lined paper or blank paper, violet paper or yellow, expensive bond or newsprint.[51]

Whereas most literary forms have evolved codified rules and conventions, in a diary anything goes; like American life, the journal can accommodate a "blindfold battle royal." Freedom and even a kind of lawlessness ("You cannot do it wrong") are also at issue here. With a few significant exceptions Rainer's words describe all the detritus—photographs, newspaper clips, letters, quotations, drawings—Lyman Ward packs into his twentieth-century daybook from the leavings of his nineteenth-century grandparents. As any of the novel's readers can attest, in the space of 565 pages Lyman alternately curses, brags, exaggerates, complains, and generally performs like a one-man Greek chorus when it comes to providing comments on the lives he creates and recreates. Lyman also defies literary conventions in numerous other ways by changing points of view and even his own style from one page to the next. Throughout his late novels, Stegner inserts passages of German, Danish, Latin, Italian, French, and Spanish, but in *Angle of Repose* he also takes care to contrast variants of Susan Burling's Quaker English against the Spanglish, Itinglish, Chinenglish, and hybrid English of minor characters like the Cousin Jacks. At times, Lyman does in fact communicate "poetically, eloquently"; at other times, "angrily, lovingly." So do Bruce Mason, Joe Allston, and Larry Morgan.

In *The New Diary* Tristine Rainer writes of the journal in terms that sometimes echo Stegner's description of wilderness. She calls the journal a "refuge" from "literary expectation and restrictions," "a private sanctuary," an "uncharted territory."[52] Just as the wilderness experience, according to Stegner, can bring "incomparable sanity" into our insane lives by providing us vacation and rest, so too the journal (and by extension the act of writing itself) can offer us "a sanctuary where all the disparate elements of a life—feelings, thoughts, dreams, hopes, fears, fantasies, practicalities, worries, fact, and intuitions—can merge" to provide a mental traveler like Joe Allston or Lyman Ward with "a sense of wholeness and coherence."[53] For Rainer a journal is "a spiritual island" full of its own unique "wilderness beauty."[54]

This correlation between our nation's few remaining islands of wilderness and Rainier's "spiritual island" (journal or log or daybook or diary) might seem contrived at first. But the connections between them point so obviously forward and backward to prototypically American experiences that we ignore the connections at our peril. Since 1492 the journal has constituted the genre of necessity for all actual voyages of discovery and exploration of "new" worlds. Much of what we know of primitive America we have learned from reading the logbooks, ledgers, and accounts of star voyageurs of outerspace like Columbus and Cabeza de Vaca, as well as from studies of the oral narratives handed down through generations by Native Americans.[55] In this sense, Lyman Ward is a consummate time traveler who marries the oral tradition to the written tradition.[56] In descriptive terms, the architecture of *Angle of Repose* is less synthetic (artificial) than eclectic: Lyman intuitively chooses what appears to be the best artifact from diverse sources. He incorporates into his narrative everything from family gossip and hearsay to his own dreams. As a storyteller, he pretty much lives up to William Stafford's definition of a creative writer: "A writer is not so much someone who has something to say as he is someone who has found a process that will bring about new things he would not have thought of if he had not started to say them."[57] Like Stegner (and Picasso), Lyman does not seek, he finds.[58] In *Jerusalem* William Blake hints at this process in these lines from "To the Christians":

> I give you the end of a golden string,
> Only wind it into a ball,
> It will lead you in at Heaven's gate
> Built in Jerusalem's wall.[59]

Indeed, Lyman's story begins by metaphorically leading us in at "Heaven's gate": "So tonight I can sit here with the tape recorder whirring no more noisily than electrified time, and say into the microphone the place and date of a sort of beginning and a sort of return: Zodiac Cottage, Grass Valley, California, April 12, 1970." If, as readers, we are "within certain fluid boundaries of general meaning and feeling" entitled to anything we can find in the writer,[60] then I choose the house of fiction, which, as Henry James pointed out, has not one window but a thousand. In these opening lines Wallace Stegner is creating his storyteller Lyman Ward as a placed person in his ancestral home, a home he must build into a place-world by telling the origin stories and history of his ancestors. By having Lyman speak his story rather than write it, Stegner is also, whether consciously or unconsciously, flirting with the Native American oral tradition. Certainly Lyman's position is similar to that at the end of Whitman's open road:

> Facing west from California's shores,
> Inquiring, tireless, seeking what is yet unfound,
> I, a child, very old, over waves, towards the house of maternity, the land of migrations, look afar,
> Look off the shores of my Western sea, the circle almost circled;
> For starting westward from Hindustan, from the vales of Kashmere,
> From Asia, from the north, from the God, the sage, and the hero,
> From the south, from the flowery peninsulas and the spice islands,
> Long having wandered since, round the earth having wander'd,
> Now I face home again, very pleas'd and joyous,
> (But where is what I started for so long ago?
> And why is it yet unfound?)[61]

Seen from this temporary angle of repose, the physical "Zodiac Cottage" in "Grass Valley" that Lyman Ward names simultaneously references (1) his ancestral home, (2) the Old English *bon hus* (or "bone house") of the human body (the dwelling place of the spirit), and (3) the Native American "earth house hold" in orbit and at play in the starry cosmos of the universe. Each body is a simultaneously discrete and yet integral part of the other and therefore dependent on—or, we might say, responsive and even relative (kindred) and responsible to—the other. From here it is not very far to Stegner's idea of *The American West as Living Space* (1987) or to his notion of the West as a paradigm of the nation and the world, "the United States, only more so," America at its most "politically reactionary, exploitive," "rootless, culturally half-baked," and "energetic" end.[62]

Likewise, as his surname suggests, Lyman is a ward (from the Old English *weard*). He is both watcher and watched, protector and protected, orphan but also guardian and custodian: the gatekeeper to the past and the present (both personal and collective). He is also the steward of what Snyder calls the writer's own "conscious agenda-planning ego" but a caretaker, too, of the nation's ego and therefore its conscience (from the Latin, *conscire*, to know wrong). In Greek mythology, Charon is the ferryman of Hades on the river Styx, but Lyman's occupation, no less than Charon's, involves the conducting of souls between worlds. Like the shapeshifter he claims to be, Lyman lives much of the time in the ghost minds and bodies of his grandparents. Obviously, in choosing to amputate the leg of his narrator, Stegner is also playing with the American literary tradition, transplanting "the hermetic grotesques of Eudora Welty and Carson McCullers and Flannery O'Connor" (and Hawthorne and Melville) all the way across the continent to show us how "western society and the western individual" are "entangled" and how all the various "Wests share a common guilt for crimes against the land that is only less bitter than the guilt of the nation for crimes against the black race,"[63] not to mention crimes against those other minorities—Indians, Hispanics, Chinese-Americans, Japanese-Americans—that Stegner chronicled in *One Nation*.

Lyman Ward's story dramatizes Stegner's belief that the body of the nation is in the mind of the nation. And here once more, I repeat, they are both wild—not alien but kindred—grotesquely maimed but not beyond recognition or recovery, restoration, and healing. "Wilderness may temporarily dwindle," Gary Snyder writes, "but wildness won't go away. A ghost wilderness hovers around the entire planet: the millions of tiny seeds of the original vegetation are hiding in the mud on the foot of the arctic tern, in the dry sands, or in the wind."[64] Although vanishing at exponential rates across the globe, vast reaches of wild desert and prairie and forest and ocean as Stegner describes in his novels still do, in fact, exist. But as we all know, the wilderness idea and the wilderness experience today stand in unprecedented jeopardy:

> Something will have gone out of us as a people if we ever let the remaining wilderness be destroyed; if we permit the last virgin forests to be turned into comic books and plastic cigarette cases; if we drive the few remaining members of wild species into zoos or to extinction; if we pollute the last clean air and dirty the last clean streams and push our paved roads through the last of the silence, so that never again will Americans be free in their own country from the noise, the exhausts, the stinks of human and automotive waste. And so that never again can we have the chance to see ourselves single, separate, vertical and individual in the world, part of the environment of trees and rocks and soil, brother to the other animals, part of the natural world and competent to belong in it. Without any remaining wilderness we are committed wholly, without chance for even momentary reflections and rest, to a headlong drive into our technological termite-life, the Brave New World of a completely man-controlled environment.[65]

In one of the most famous of the many famous phrases he coined, Stegner refers to wilderness as "a part of the geography of hope,"[66] a phrase echoing Thoreau's original pronouncement: "In wilderness is the preservation of the world." On this level the issue of wilderness preservation ultimately becomes a matter of self-interest for the human species. Pollute the pond and the lilies die. Destroy the earth and we destroy ourselves. But the destruction of wilderness, Stegner argues, leads also to perversions of identity. It leads away from freedom and individuality toward rigid conformity and Eigil Rodding's incestuous and evil island empire: to mass distortions of character—insanity and madness. But for those who value deity, there is even more at issue than the relationship between the body physical and the body politic. In Stegner's view, wilderness takes us as close as we can get to "whatever God" we choose to believe in.[67] In other words, Stegner's view of wilderness inextricably involves us in sacred matters. The destruction of the earth constitutes no less than blasphemy: flinging God's gifts in His face.

Most of us have a long way to go before we catch up to Wallace Stegner, whose ambition it was to become a "good animal." At the conclusion of our formal interviews Stegner asked: "Haven't we agreed with Frost that there are no new ways to be new?" No new ways, then,

> only reiterated and intensified versions of the old ones: the obligation to use oneself to the bone, to be as good as one's endowments and circumstances let one be, to project one's actions over and beyond the personal. The only things I owe to myself I owe to my notions of justice. But I owe a great deal, in the way not only of obligation but of tenderness, to my family and my friends. Chekhov said he worked all his life to get the slave out of himself. I guess I feel my obligation is to get the selfishness and greed, which often translates as the Americanism, out of myself. I want to be a citizen of the culture, of the best the culture stands for, not of a nation or a party or an economic system.[68]

Across a literary career that spanned half a century and an even longer lifetime of experience, Wallace Stegner worked strenuously to become a new kind of American. Consequently, he also became a new kind of American writer, one who learned ways to protect and live with the earth and teach others. He knew who he was because he knew where he was. His farsighted, continental vision, fused from self-knowledge and the knowledge of his region and his culture, carries much of the power we need to generate the future. If we use it wisely, it can help us steer a civilization that threatens always to speed out of control, even as it nears the cliffs of extinction.

PART II

STEGNER AS HISTORIAN

As a historian Stegner was cut from unorthodox but durable cloth. In Stegner's experience the historical—once he discovered it—was always in large part the personal. He gained his historical perspective only with experience; it was bound closely to his own evolving sense of self. Stegner recalled growing up in the newly "opened" province of Saskatchewan: "The general assumption of all of us, child or adult, was that this was a new country and that a new country had no history. History was something that applied to other places."[1] One may trace much of the growth of Stegner's continental vision through his overturning of this assumption. The personal nature of that experience, and that struggle toward understanding, made Stegner's histories more than well-wrought evocations of the past; it made them continuous with the present and the future and with Stegner's other literary genres.

At the core of Stegner's work as a historian is his effort to understand the mythology of the American West and its patterns of development. As Elliott West observes, "No one has argued more passionately and persuasively [than Stegner] for a unified reconstruction of western history." Since the early 1980s his lead (as well as those of Walter Prescott Webb, Bernard De Voto, and others) has been taken up by a generation of "New Western Historians" who are reinterpreting the history of the American West from a broader range of perspectives. Historians in this movement have focused on the diverse cultural influences that have shaped the contemporary West but that have often been excluded or marginalized in the standard historical narratives. During this same period, environmental history, which overlaps substantially with the New Western History, has also emerged as a field of its own, focusing on the dynamic interactions of culture and nature. Stegner may be considered a forerunner in both areas. He composed important reinterpretations not only of regional history (in, for example, *Mormon Country*, *Beyond the Hundredth Meridian*, and *The Gathering of Zion*) but also regional environments (in, for example, *Wolf Willow*, *The Sound of Mountain Water*, and *American Places*). His appreciation of a broadened sense of western history is, of course, also woven into his fictions, especially *The Big Rock Candy Mountain* and *Angle of Repose*.

The essays in this section cover Stegner's contributions as a historian. Elliott West pays Stegner the compliment, not merely of lauding his "remarkable encompassing vision of the West, present and past," but of critically examining "a striking anomaly" within it: Stegner's essentially traditional view of wilderness as static and disconnected in time and place and (inlaid, in a sense, within this view) his failure to integrate the experience of Native Americans more fully into his work. Lending a historian's perspective to the effort that James Hepworth began in Part I, West finds lessons in Stegner's deficiency. "Like everything he did," West writes, "even this flaw is provocative and full of potential for a deeper understanding."

Walter Nugent places Stegner himself in historical context by asking how and why Stegner's definition of the West changed, and what we might learn from this. Nugent concludes that "despite many continuities, [Stegner's] views became a little less rigorous and consistent" and "his definition of the West shifted," drawing on both romantic and realistic traditions. The flux—between the open West and the urban West, between a constricted past and an uncertain future—is not, of course, unique to Stegner. As Nugent observes, "The demythologizing continues. So, too, however, does a romanticized West." In the contemporary West, changing at breakneck pace, this tension painfully frames the issues of the day. For Stegner, it also framed history and his response to it. In the end, he found his optimism increasingly at odds with his realism, yet still seeking to reconcile various Wests.

Dan Flores asks us to step back and see Stegner as part of a longer process that began "with Jefferson on his mountain with the Blue Ridge lining the horizon"—a process that is bound to continue until we get it right: the process whereby Euro-Americans may yet "become native to this place."[2] Flores sees Stegner as "a transition figure in the tradition of the great American literary environmentalists of place" and "a proto-bioregional citizen of the larger West." Stegner inherited and extended a tradition that includes Henry David Thoreau, John Muir, Mary Austin, Aldo Leopold, and Rachel Carson. In adapting that tradition to the West's distinctive regional features and infusing it with his impressive grasp of history, Stegner created a new and consistent "critical vision of the West," a vision that a growing number of people, writers and citizens alike, now draw upon and continue to formulate.

Stegner's two major biographies—of John Wesley Powell and Bernard De Voto—functioned not only as works of history but also as studies in the relationships among character, society, and landscape. Curt Meine's essay examines Stegner's biographical works, noting that they are "distinguished by their examination of the relationship between biographical subject and biogeographical space and by the way Stegner uses this relationship to examine forces, tensions, patterns, and themes at the heart of America's cultural development." More directly perhaps than other genres, biography asks how the interior and exterior interpenetrate in the lives of human beings. By expanding the genre to encompass the environment, Stegner created new possibilities for the art and new questions for its practitioners.

Stegner held that his interest in history was "personal, not scholarly. . . . All the history and biography I've done has been an offshoot of personal experiences and personal acquaintances."[3] But of course there was no clear line between the personal and the scholarly. Throughout his career Stegner occupied the precarious middle ground between literature and history. As Gary Topping has shown, Stegner's approach to history entailed at least two of the professional historian's essential qualities: mastery of sources and the passionate belief that history matters, personally and politically. To these, however, Stegner added the writer's eye for symbol and ear for drama; he invariably brought to his history the "literary impulse . . . [to elicit] emotional response."[4] This combination of qualities left him open to censure, but it also allowed him to explore the unique potential in his voice. And when combined with his characteristic sense of personal responsibility, it finally led him not only to write but to act.

CHAPTER SIX

WALLACE STEGNER'S WEST, WILDERNESS, AND HISTORY

ELLIOTT WEST

THROUGH HIS LONG CAREER, Wallace Stegner strove toward a unified vision of the American West. He moved along various forms of expression, blurring the boundaries among fiction, history, memoir, and essay. More than most writers, he refused to recognize any border between the author and his work; his novels as well as his nonfiction commentaries are woven together tightly with his personal beliefs. Over and over he called for an end to the disastrous alienation of westerners from the land he considered a treasured birthright. And no one has argued more passionately and persuasively for a unified reconstruction of western history. Above all Stegner spoke to that deep fracture in the usual perception of the western past—a rift separating the present-day from the imagined West of the nineteenth-century frontier.

This desire to unify—to reach across and heal that historical divide—is evident in many of Stegner's fictional creations. Consider Bruce Mason at the end of *The Big Rock Candy Mountain*, standing beside his father's grave, hoping to reconcile with a remembered past and fulfill in his own future the best promise of his parents' lives. Consider Lyman Ward, the western historian in *Angle of Repose*, who sits in his wheelchair, one leg gone and the other useless, as he sifts through his grandparents' papers at the novel's beginning. Piecing together their lives, he sees that he is trying as well to place himself in his family's story and, through that, to situate himself in his immediate world. "As I look down my nose to where my left leg bends and my right leg stops," Ward tells us, "I realize that it isn't backward that I want to go but downward. I want to touch once more the ground that I have been maimed away from."[1]

In several essays Stegner describes modern westerners as experiencing a predicament much like Lyman Ward's. They have been cut off from their past and therefore from a genuine connection to the country they inhabit, the land they walk on. Western literature, obsessed with the pioneer tribulations and "horseback virtues" of the past century, has told a story centered on its own heroism and on events that supposedly ended with a gigantic clang around the opening of the twentieth century. That story cannot build into any kind of future. And so what should have been its future—our present—is marooned in the here-and-now with no awareness of where it has come from. "Millions of westerners, old and new," Stegner wrote thirty years ago, "have no sense of a personal and *possessed* past, no sense of any continuity between the real western past which has been mythicized almost out of recognizability and the real western present that seems as cut-off and pointless as a ride on a merry-go-round that can't be stopped."[2]

The job before us, he said, is somehow to close that gap between the pioneer and the modern—to retell the western story so it moves unbroken, with only the normal rough spots, from then until now. We need (literally) to re-member what is now a dis-membered past. Such a remembering, Stegner thought, will promote a more sensible, compassionate approach toward western lands. If westerners feel properly grounded in a continuous past that is theirs alone, they will see how their unique story has unfolded in the land and shaped the land they see around them. Good history feeds a more rightful treatment of the country we live in—not in the generic, sappy, new age sense, but as an awareness of our natural neighborhoods as a part of ourselves and of the others who have woven this history of seamless generations over the past century and a half. We don't have to think of our immediate world as Mother Earth and Father Sky. Maybe it's more like old Aunt Minnie, who talks too loud and crunches ice at the dinner table, or a deadbeat stepbrother who avoids shampoo and runs up the phone bill. We will not always like how our places behave and treat us, but they're kin. We accept them, and we try to live with them respectfully and with some loyalty.

This was Stegner's gift to us, I think—an expression of the West's better possibilities and a personal demonstration of how they might be achieved, the personal example of fifty years of hard work well done, in novels and short stories, in history and biography and essays, in memoir, and in public advocacy. The theme throughout was unity: a convergence of these different expressions and, within each, a healthy tendency toward completeness.

But Stegner's accomplishment, for all its scope, is in its own way inconsistent. His remarkable encompassing vision of the West, present and past, has within it a striking anomaly that I think is especially revealing. To me, at least, this makes him more interesting. Like everything he did, even this flaw is provocative and full of potential for a deeper understanding. Just as Stegner's insights run easily and unbroken through his several modes of expression, so too this anomaly can be found in his histories, his fiction, his essays on conservation, and his thoughts about himself. Like his insights, it forms a kind of loop through his body of writing. We can break into that loop by looking at one idea that appears often in Stegner's work—the idea of "wilderness."

In contrast to so much else in his work, Stegner's perception of wilderness is in many ways remarkably traditional. It is rooted in the idea of a clear and distinct division between an advancing human influence and a natural world virtually free of human touch. It is a variation of the dream of finding that land from before the fall, before the corruptions that human society brings to nature's perfection. In 1990 Stegner quoted approvingly George Perkins Marsh on what it meant when society arrived in the wilderness. "Man," Marsh wrote, "is everywhere a disturbing agent. Wherever he plants his foot, the harmonies of nature are turned to discords."[3] When Stegner found parts of the West he believed were still beyond the reach of that influence, still tuned to nature's harmony, he could make this mythic reference explicit. He titled his 1947 account of visiting Havasu Canyon "Packhorse Paradise." More than forty years later he recalled a trek to a beautiful spot high in the Uintas, with sheltering trees, etherlike air, and trout that all but jump into your skillet. He called this short essay "Crossing to Eden."

Surely Stegner's most famous summation on the subject is found in his "Wilderness Letter" written to David Pesonen in 1960. Here he argues that wild areas should be protected not only for their potential recreational use but, just as important, for "the wilderness *idea,*" which he calls "an intangible and spiritual resource." Americans just need to know that there are parts of our land still beyond modern society's discordant presence. "We simply need that wild country available to us," Stegner wrote in the oft-quoted conclusion of the letter, "even if we never do more than drive to its edge and look in. For it can be a means of reassuring ourselves of our sanity as creatures, a part of the geography of hope."

We can all sympathize with this notion, but we don't have to look too closely to see a striking difference between this and so much else that Stegner wrote. If much of his work is about continuities, connections, and unity, the references here are to divisions, separate worlds, lines, and edges. The West described here is not of a piece, but a West divided into starkly different domains. There is human society and there is the wilderness; there is the built and the wild.

This dichotomy raises problems. One practical difficulty involves the good effort to protect land that remains relatively free of twentieth-century technology's impacts. Setting aside such land, then treating it as "wilderness," has not had the desired effects. The most obvious instances are those national parks originally created as islands of nature to be kept "unimpeded by man." In preserving "wild" ecosystems, there is an elusive goal. In Rocky Mountain National Park, for example, administrators might let natural fires burn as they will within the park, but development and fire control just outside the line interrupt the usual pattern of low-altitude blazes spreading upward—so when fire does come to the higher, cooler, wetter country, it burns far fiercer than in the past. When the system's superpredators (grizzlies, wolves, cougars) are killed or squeezed out by human sprawl just over the line, their prey proliferate. Elk overgraze the willows and young aspen, denying beavers food and materials to build dams; the decline of dams alters old patterns of water flow, speeding erosion and draining the wetlands, which further undercuts the diverse floral cover. Between 1940 and 1980, by one estimate, the beaver population in the Big Thompson drainage fell from 315 to 12.[4] In the meantime, outside the park, where wilderness supposedly has been banished forever, yuppie suburbanites have been busily planting trees. Happy, full-bellied beavers are multiplying there like fifty-pound gerbils.

At work here is a contradiction present from the earliest attempts to protect parts of the West from the full brunt of modern change. On the one hand preservationists recognized the principle of unity and in fact could use it to stress the public's kinship to the nonhuman. "We all travel the Milky Way together, trees and men," John Muir wrote. "When we try to pick out anything by itself, we find it hitched to everything else in the universe."[5] But in the next breath Muir and others would use this same appeal to support policies that stood in startling contradiction to the argument itself. The public was reminded of the seamless connectedness of things—and then promised places starkly separate and seemingly disconnected from the land around them, islands of the changeless within a land of dynamic change.

All this might seem to belong more in the conservation section of this volume, but I think that Stegner's view of wilderness also says something important, interesting, and somewhat troubling about his view of western history. The idea of wilderness—as country separated cleanly by a firm line from the developed West—corresponds to a certain view of western chronology. According to this view, history doesn't really get going until Europeans show up and start changing things. Before that, this view further suggests, this region was strangely static, remarkably free of human influence. Today's wilderness is wilderness—it is what it is—because, by some lucky convergence of circumstances, it supposedly has managed to stay largely beyond the reach of human change. Standing on the boundary where wilderness begins, we can do more than gaze into a wild, magnificent landscape. When we drive to that line, to paraphrase Stegner's "Wilderness Letter," we can see America still untouched by events that set our collective story in motion. We are looking over the edge of history itself.

This perspective, of course, raises the question of the Indian presence in the landscape. If the land beyond the line is a residue of country that was untouched by human history when Europeans arrived, then who were those people living there when the Spanish, French, and English showed up, and what in the world had they been doing all those centuries? Especially when arguing for a saner treatment of the land, Stegner slipped often into imagery of Europeans moving into a continent virtually devoid of human presence. The "salient historical fact about the West," he wrote in "Born a Square," was "the confrontation between empty land and imported populations."[6] Western history, he said in a 1978 interview, has been the story of a "high energy civilization investing and changing and virtually destroying a virgin continent."[7] The continent, he wrote in an essay for the Wisconsin Humanities Committee, once "stretched away westward without names. It had no places in it until people had named them."[8] And when did the naming start? As an example he cites Kentucky, which became a place only when Daniel Boone remembered a hunting spot as Bear Run and other settlers picked up the name.

In other writing Stegner did recognize ancient, vanished peoples but none who seemed to qualify as part of history. The Latter-day Saints sought their desert refuge in an area that was home long ago to the Ho-Ho-Kam, he wrote in an early chapter of *Mormon Country*. But at the time the Mormons took possession, in the 1840s, he gives not the slightest hint that anyone at all was living in the Great Basin. He called this chapter "The Land Nobody Wanted." Some Native Americans were hard to ignore, however. Havasu Canyon, the "Packhorse Paradise" Stegner visited in the 1940s, was of course home to the Havasupai. But he described them in historically flat terms. What makes them Indians, in fact, is their unchanging nature and their lack of dynamic, evolving interaction with people and land around them. They lived a "static life," he wrote, with its "dynamics . . . reduced to the simple repetition of simple routine, its needs few and its speculations uncomplicated."[9]

There is not much room in this view of Indians for historical and cultural evolution, especially in their exchanges with European Americans. In *Beyond the Hundredth Meridian*, Stegner wrote that Indians have failed, not in their physical survival, but in continuing their tribal cultures. Even in resistant, isolated southwestern groups buffered from change by aridity, he found signs of decay: in a dancing Hopi who wears an Ingersoll watch or Purple Heart as well as tortoiseshell and turquoise, and among Navajo women who chew bubblegum and wear saddle shoes and sunglasses while herding sheep. These, he wrote, are indications of a coming cultural extinction. The point in these passages seems to be that change equals the death of Native America. An Indian with a wristwatch is surrendering his or her culture, because that culture is not deemed to be part of a legitimate historical process of evolution and adaptation. Indians are by definition part of a world in stasis; they live on the other side of the line between history and paradise, the line we cross into Eden.

This odd treatment of Indians follows from the contradiction, noted above, in Stegner's chronology. Stegner argued that our heroic view of the nineteenth-century West prevents us from connecting that earlier mythic time to the West we know today. True enough. But Stegner also pictured a similar disruption on the other end, the far side, of the pioneer West. His mythic view of the West before Europeans—and the static wilderness world in which Indians supposedly lived—creates a great unbridgeable break at the start of the frontier very much like the one Stegner so deplored at the frontier's end. It allows no sense of continuity between the human story on the other side of that break—the story of ancient native peoples—and the story on our side. Consequently, Stegner seems unsure just what to do with these people, the Indians, who have managed to live across that chronological chasm. They are ignored altogether, or they appear as vanished residents of a distant past, or they are static holdovers from that timeless wilderness world.

At this point we must step back and ask: was Wallace Stegner largely unaware of Indians and their place in the western story? Of course not. He wrote something of them (and their names on the land) in *Beyond the Hundredth Meridian*. Two years after "Packhorse Paradise" he published a short piece, "Navajo Rodeo," a concise description of one bit of Native American life. The remarkable but seldom mentioned book, *One Nation*, a collection of photographs from *Look Magazine* for which Stegner wrote the text, includes a section on Indians (mostly southwestern and plains) in 1945 America.

It is fair to say, I think, that his knowledge of Native America did not run nearly so deep as in most other aspects of western history and life. But the question is not really how much he knew but how he used what he did know. He was plenty aware of Indians, but he used that knowledge, not in service of the unified vision so evident in his other work, but to picture a West fractured historically and divided within itself. And that, I think, stands in the way of Stegner's larger aspiration—the prospect of a more accurate and realistic telling of western history, and through that truer telling a western society more fully connected, in identity and obligation, to western places.

All this should encourage us to build on what Stegner began. His work suggests to me three useful lessons for western historians today. In fact, we can read Stegner himself as calling us to this work. In 1986, in "The Sense of Place," Stegner wrote that "we need to know our history in much greater depth." And so we do. We ought to recognize that the story of human inhabition of the West is very old—and we should follow the implications. So the first lesson is that we need to go deeper, drawing far more from the fields of anthropology—not so much in its applications to contemporary cultures but in its revelations about much earlier peoples—and especially from archaeology. We need at least to sketch in earlier patterns of change and development and how they meshed into that part of the story that opened with the European intrusion.

The evidence is there, often staring us in the face. On the boundary between Rocky Mountain National Park and the resort town of Estes Park is a cone-like formation of rather barren stone known locally as Old Man Mountain. Recent digs suggest that this was a ritual site for at least three thousand years before whites saw it. Standing as it does on the newly imposed line between the developed resort and the unhistoried park, it is a healthy reminder of the dangers of separating the recent from an earlier past. When Moses was on Sinai, native peoples were worshiping here, looking down on land now occupied by a Seven-Eleven and the Jolly Jug Liquor Store.

Recognizing this antiquity leads us to the second lesson. It encourages us to take a closer look at words we have taken far too lightly: "wilderness" and "wild." A lot has been written lately about the "wilderness idea" and the complex cultural assumptions behind it. It might help to push this discussion forward by suggesting that "wilderness" and "wild" have carried with them two quite different meanings. In much of the writing about the West, Stegner's included, those meanings are often balled up together. The results have been unfortunate.

On the one hand, "wilderness" can mean the expanse of the nonhuman, the natural world's glorious snarl that exists with us but apart from us. Defined this way, wilderness can be found in many places. But certain examples of it—the Salmon River watershed of central Idaho, for instance, or the canyon country of southern Utah—are especially valuable to us because encountering them is such an overwhelming experience that their meanings and lessons are nearly impossible to ignore. They are also incomparably beautiful. Stegner insisted that "wilderness" by this definition was something we must protect, learn from, and cherish. Such country also reminds us of those parts of our individual beings, our instincts and drives, that we have in common with other creatures. These wildlands teach us that no matter how hard we try to fool ourselves with technology and cant, we are ultimately inseparable, inside and out, from God's other work. We cannot possibly understand ourselves and our history without seeing human society as always bound up in its nonhuman setting, whether this wilderness is of redwoods or crabgrass, and we cannot learn to be "good animals," to use Stegner's term, without first recognizing the animal in each of us.

But "wilderness" is often given a second meaning—country that is almost completely free of human presence, now and in the past. This definition admits that these places may have some limited human experience, but it denies that such experience adds up to anything we recognize as history. In both the distant and more recent past the people of this wilderness are depicted as having led lives that were largely static over long stretches of time. Looking back, this notion seems almost indistinguishable from another dubious perception—that of an unchanging physical world. This view of things at least avoids the problem of drawing artificial lines between these earlier native peoples and the "wild" land around them, lines we have drawn for our modern westerners. In fact, ancient Native Americans are often pictured as variations of exotic fauna. Like a scorpion in amber on a clip of a westerner's bolo tie, they are figures frozen in time.

These two meanings of "the wild" have been inextricably woven together. Certainly that was true of Stegner's work. We tend to assume without thinking that embracing one necessarily requires accepting the other. But that is not the case. One role of historians today might be to move beyond the present discussion of the "wilderness idea" by distinguishing between these varied definitions and then exploring their implications in understanding our past and affecting our future. This much, in my opinion, is clear: we must accept the first meaning in order to live sensibly and ethically; but the second use of "wilderness" is full of potential mischief.

For one thing, using "the wild" to mean land largely without people or history muddles the important task of protecting country that is "wild" in the first sense. It leaves us working toward a goal—reestablishing an environment virtually free of human influence—that is hopelessly unattainable. This faulty view of our wilderness complicates another vital job, especially for recent arrivals in the West and for the descendants of European-American pioneers. Stegner taught us that we will never earn that elusive quality—a "sense of place"—without acknowledging, at least in spare outline, the full story of where we live, whether the Oregon desert or a Denver suburb. It follows that we cannot truly know western places without recognizing their long and complex histories prior to the coming of Europeans, any more than we can develop an intimate friendship with someone who refuses to tell us anything that happened to him before last week.

And that implies something even more troubling. If wilderness is Edenic country largely free of the human story, then some of today's westerners, those descended from the region's first inhabitants, are cut off from their narrative roots. A continental vision that fails to recognize North America's extraordinary history before 1500 will deny Native Americans what they need for their own distinctive understanding of who and where they are now. They will be as isolated from their origins as those other modern-day westerners, described so well by Stegner, who have been marooned in the present by the botched historical view of the pioneer West. What a bleak joke it would be if European Americans come to believe that they can find their "sense of place" only by denying to their neighbors what *they* must have for their own legitimate sense of belonging.

So the first two lessons implied in Stegner's legacy—exploring more fully the West's antiquity and taking a closer look at our ideas of "wilderness"—lead naturally to a third. We ought to give more prominence in western history to those with the longest lineage in that old, tangled story: the Indians. We might begin by rethinking some of Stegner's portrayals (and nonportrayals) of native peoples. The Great Basin, of course, was not a "Land Nobody Wanted" when the Mormons arrived. Shoshonean and Paiute peoples were there. They wanted it, but they lost outright control. They also survived and adapted, as they had for dozens of generations. They joined with the new arrivals, were exploited by them, and in their own way they used the Mormons. They produced, among other things, the native revivalist Ghost Dance religions of 1870 and 1889, both influenced by LDS theology (and neither mentioned in *Mormon Country*). The Havasupai still live in Stegner's "Packhorse Paradise." They don't live especially simple lives of uncomplicated speculations, but then they never did. They continue to interact with the world around them in various ways. In its promotional pamphlet, the nearby Enchantment Resort boasts the world's only full-blooded Havasupai concierge.

Integrating Indians more realistically into the West's long history will help in other tasks that Stegner considered essential. Seen as country layered deep in experience, this new "wild West" becomes a far better guide for respectful treatment of beloved country. Its story is a great catalog of lessons from a hundred and twenty centuries of exchanges among people, animals, soil, plants, and landforms. Newcomers and seasoned westerns alike will be more likely to understand their country and respect it more deeply when the long native story becomes part of our common narrative.

"Plunging into a future through a landscape that had no history," Stegner wrote, "we did both the country and ourselves some harm along with some good." Neither we nor the places where we live, he added, can ever be healthy until we "acquire some sense not of ownership but of belonging."[10] Though the landscape he referred to was a twentieth-century West severed from its true pioneer origins, we can take Stegner at his essential word and carry his good insight further. We and our region will be far healthier if our continental vision embraces all western peoples in the full depth of their pasts. We will be closer to that sense of belonging when we recognize that the coming of Europeans was not an occupation of a wilderness without history but a participation, granted a wrenching and disrupting one, in an old story of migration, settlement, conflict, movement, use and misuse, error, adaptation, and survival.

It is in the experiences of Indian peoples, I think, that we are likely to find some vital, strong foundations for the unified understanding of the West that Stegner called for so eloquently. And as westerners begin to anchor their history in real places with a continuous past, Indians will begin to assume the roles they ought rightfully to play. A fine example comes, ironically, from Stegner himself. Of the many intriguing points about this man, one of the most fascinating is this: in all his large body of work, only once does he bring Indians into his narrative as flesh-and-blood humans with long pasts, evolving identities, and, at least by implication, some kind of future. And this happens on the one occasion when Stegner writes most personally about that one place where his own character was truly rooted, where he came closest to that "sense of belonging." The place, of course, was around Whitemud and Eastend, in southern Saskatchewan, where Stegner spent his fifth to eleventh years. Those, he wrote later, "were the shaping years of my life. I have never forgotten a detail of them."[11] The work was *Wolf Willow*, part memoir, part history, part speculation, all of it an intimate engagement with his own past. In *Wolf Willow* we meet the Métis, mixed blood of French, Algonquin, and Huron, who have grown out of Whitemud's past to play a complex and ambiguous role in its story. In other words, it was in precisely that place closest to Stegner's identity—that place where he himself felt what he wanted for all westerners: a "personal and possessed past"—it was there that Indians suddenly appear, not as relics frozen in the past, but moving naturally through the story, full participants in an ongoing history.

Wolf Willow suggests the promise to be gained from following Stegner's lead and building on his enormous contribution by learning from his failings. A western story that "does not take into account time and change," he wrote in 1967, "can only give us a fractured view of our history and ourselves."[12] He was referring once again to the disjunction between our century and the one just before it, but his words apply just as well to that earlier break in western history, the one between the long history of native peoples and the coming of the Europeans. Both need healing. If, as the new historians continually remind us, the region's history did not end with the populists, neither did it begin with Cabeza de Vaca. By bridging that earlier break, we can work toward a fuller appreciation of our land, a saner means of living with it, and a fairer reckoning with all its people.

In his "Wilderness Letter" Stegner gives us a memorable image of driving to the edge of wilderness and looking in. In one of his best essays, "History, Myth, and the Western Writer," he uses an image at least as vivid. Arguing the need to build narrative connections between the West of today and yesterday, so that each might serve the other, he recalls how during the worst blizzards pioneer settlers would tie lariats together to stretch between house and barn to find their way back and forth. "With personal, family, and cultural chores to do," he writes, "I think we had better rig up such a line between past and present."[13] But we'll need a lot of lariats, many more than Stegner had in mind, and we'll need to rig up lines that are very long indeed.

Stegner's work is an extraordinary gift, but I suspect he would want us to see it as an obligation as well. We should think of it as an affluence—in both definitions of that word. It is a great richness, an abundance that we should celebrate. And, in the original meaning of affluence, it is also a flowing toward something even larger, a generous stream that, in the logic of its advance, points us in the direction we need to go. If we move with it, and learn all we can from this remarkable man, we will know our place and its people and ourselves much better, and perhaps the West will become, in a different but truer way, what Stegner said it is: a geography of hope.

CHAPTER SEVEN

WALLACE STEGNER, JOHN WESLEY POWELL, AND THE SHRINKING WEST

WALTER NUGENT

WHAT DID "THE WEST" MEAN to Wallace Stegner? Did his definition change with time and experience? Can we figure out why? Interesting as such inquiries into Stegner's mind may be, it is probably even more important to consider how his definitions—right, wrong, or partial—can help us define and understand the West. He was after all not just a thoughtful man but a provocative one. My guess is that he would not be satisfied unless we learned not only what he had in mind but where we ought to take it on our own.

I don't think it demeans Stegner to call him a western writer, nor would he have thought so. There is nothing wrong with being a "western writer"—especially if, as I think is true, the West has become the most dynamic American region in the late twentieth century and promises to remain so into the twenty-first. But was Stegner a western historian as well as a western writer? My impression had always been that Stegner had written many novels but only one work of history, *Beyond the Hundredth Meridian: John Wesley Powell and the Second Opening of the West*, which came out in 1954. It is a book I and many others have always considered to be a first-rate work of history, even though it purports in some respects to be a biography. But Stegner did more history than that. If we place our trust in the catalogers at the Library of Congress, we find that Stegner's works classified as "E" or "F," American general or regional history, appeared throughout his life. They include *Mormon Country* (1942); *One Nation* (1945); *Wolf Willow: A History, a Story, and a Memory of the Last Plains Frontier* (1962); *The Gathering of Zion: The Story of the Mormon Trail* (1964); *American Places: Human, Natural, and Personal History* (1981); *The American West as Living Space* (1987); as well as *Hundredth Meridian*. Beyond all of these, he reflected often on western history, as in his biography of Bernard De Voto, *The Uneasy Chair* (1974); his extended interview with Richard W. Etulain, *Conversations with Wallace Stegner on Western History and Literature* (1983); and his final book, *Where the Bluebird Sings to the Lemonade Springs* (1992).

This is a rather impressive historical corpus for a mere novelist and teacher of writing. But to answer the questions initially posed—how did Stegner define the West and what can we learn from him?—I confine myself here to *Beyond the Hundredth Meridian*, the Etulain *Conversations*, and *Where the Bluebird Sings*. These yield a cross-section of his views from the early 1950s, 1980s, and 1990s. Despite many continuities, his views became a little less rigorous and consistent in the last volume compared to the earlier ones.

What then did Stegner say about the West in *Where the Bluebird Sings*? This book appeared after the "New Western History" exploded out of the shadows of western historiography into the national media, demanding more attention to the twentieth-century West, to people other than white Anglo males, and to stories that were not triumphalist, maybe not even edifying. It demanded, in short, that western history be freed from myth. If the history of the South could be freed from Tara and Scarlett by honestly confronting the racial and class conflicts in its past, then the history of the West should be freed from Owen Wister's Virginian and the Madonna of the Trail by honestly looking at what happened to Indians, Hispanics, and Asians and all the rest of its many peoples. The demythologizing continues. So too, however, does a romanticized West. Where Stegner stood, or might have stood, is not entirely clear because he was both a hard-nosed historical realist and a romantic. His definition of the West shifted, therefore, drawing on both camps.

About five years ago I surveyed several hundred historians and fiction writers whose professional subject matter, historical or literary, is the West. I asked them where they thought the West is and why they thought so. (To my great regret, I did not send a questionnaire to Stegner.) The results appear in the summer 1992 issue of *Montana: The Magazine of Western His-*

tory. Here I point out only one: a polarity between the historians and the fiction writers—between, that is, the hard-noses and the romantics. At one extreme, a western novelist who lives in Youngstown, Ohio, replied: "The West is a thing of the imagination, not of boundaries." Many others agreed. The West is not a place at all, they insisted, at least not any more, but a state of mind, a memory of a place that stopped existing around 1915. At the opposite pole, a historian replied that he defined the West solely in accord with the United States Census: by its latitude and longitude. His response was unusually laconic, but most historians in effect agreed by naming geographic boundaries. These respondents (and some of them were state-of-mind people who did so grudgingly) fell broadly into two camps: a majority believed that the West occupies the space from about the 98th or 100th meridian to the Pacific; a considerable minority limited it to the area from the High Plains—Deadwood, Dodge City, the Pecos—westward to the Sierras, excluding California (or most of it) and the metropolitan coastal and sunbelt West. We can call the first definition the inclusive West and the second the arid West. The arid West has lots of horses and few humans; the inclusive West has few horses, many cars, and big cities as well as mountains and deserts.

Where would Stegner have placed himself in this polarity? We do have his views from *Where the Bluebird Sings*, published in 1992. And in this book, as I read it, he defined the West in two different ways. In the first paragraph, he writes that from his birth in Iowa in 1909 until September 1930, "when I boarded a bus in Salt Lake City to go 'back East' to graduate school in Iowa, all the places I knew were western: North Dakota wheat towns, Washington logging camps, Saskatchewan prairie hamlets and lonely homesteads, and the cities of Seattle, Great Falls, Salt Lake, Hollywood, and Reno, with a lot of country seen on the fly between them."[1] Thus not only open country but Seattle, Hollywood, and Reno are parts of the West.

In beginning his second chapter, "Thoughts in a Dry Land," he defines the West very unambiguously. It "stretches from around the ninety-eighth meridian to the Pacific, and from the forty-ninth parallel to the Mexican border, [and] is actually half a dozen subregions as different from one another as the Olympic rain forest is from Utah's slickrock country or Seattle from Santa Fe" (pp. 45–46). This puts him firmly in the inclusive camp: his words are almost identical with the majority view of my questionnaire respondents. But thereafter, what was so clear begins to get cloudy. A more romantic vision begins to rise above the hardpan of geography. "The Westerner," he writes, "is less a person than a continuing adaptation. The West is less a place than a process. And the western landscape . . . has now become our most valuable natural resource, as subject to raid and ruin as the more concrete resources that have suffered from our rapacity" (p. 55). And then, as one began to suspect might happen, he redefines the West as the arid region, excluding the wet and the urban. Thus: "Actually it is not the arbitrary 98th meridian that marks the West's beginning, but a perceptible line of *real* import that roughly coincides with it. . . . This is the isohyetal line of twenty inches, beyond which the mean annual rainfall is less than the twenty inches normally necessary for unirrigated crops" (p. 59; emphasis added). Do Seattle, Hollywood, and Reno have crops? Where did they go? Are they excluded from the West because they are cropless?

Thus Stegner arrives at his second answer, and it is not the same as his first:

> So—the West that we are talking about comprises a dry core of eight public-lands states—Arizona, Colorado, Idaho, Montana, Nevada, New Mexico, Utah, and Wyoming—plus two marginal areas. The first of these is the western part of the Dakotas, Nebraska, Kansas, Oklahoma, and Texas, authentically dry but with only minimal public lands. The second is the West Coast—Washington, Oregon, and California—with extensive arid lands but with well-watered coastal strips and also many rivers. Those marginal areas I do not intend to exclude, but they do complicate statistics. If I cite figures, they will often be for the states of the dry core. [pp. 60–61]

In other words, parts of Washington, Oregon, and California still belong in the West but not the "well-watered coastal strips," where the cities are. As for cities, he writes,

> forget the Pacific Coast, furiously bent on becoming Conurbia from Portland to San Diego. Forget the metropolitan sprawl of Denver, Phoenix, Tucson, Albuquerque, Dallas–Fort Worth, and Salt Lake City, growing to the limits of their water and beyond, like bacterial cultures overflowing the edges of their agar dishes and beginning to sicken on their own wastes. If we want characteristic western towns we must look for them, paradoxically, beyond the West's prevailing urbanism, out in the boondocks where the interstates do not reach, mainline planes do not fly, and branch plants do not locate. [p. 73]

Along with Indian writer William Least Heat Moon, he finds "the true West somewhere outside the cities where 75 percent of Westerners live" (p. 75). If there is any doubt about his position, he insists that "California, it should be said, is a separate problem, hardly part of the West at all. As one anthologist recently put it, it is west of the west" (p. 138). He writes this after thirty-seven years of living in Stanford. He has taken about seventy pages to get from definition 1 to definition 2, but he finally arrives at an arid-region, nonurban, nonpeopled West—a West that is more romantic than real, more oriented to the past than to the future, and therefore less helpful than his first definition and what he wrote in earlier books. Whatever he thought in 1992 of the New Western History, he was not buying much of it.

Yet, in his *Conversations* book with Richard Etulain in 1983, Stegner seemed to anticipate some of the main thrusts of the yet-unborn New Western History. Stegner blamed the mythology about the "Old West" for the reluctance of historians to carry on into the contemporary West:

> It's not merely the Pacific Slope that's mainly urban, it's the whole West. . . . I don't know what the figure is now, but it must be eighty percent urban. [Stegner is right.] It's an oasis civilization. And curiously, that little twenty percent of rural Westerners have put their stamp on very large parts of the West, probably because of the romance of the horseman. . . . When we think of the West, you're right, we do think of it as a lonesome horseman far off on some butte. And I haven't seen very many of those.[2]

This is realism—both about the cowboy myth and about the urban nature of the twentieth-century West. Nothing in these interviews from the early 1980s, when Stegner was seventy-two or seventy-three, is at all inconsistent with this realism. He hoped, in fact, that some historian or group of historians would write a modern history of the West, a "history that tells what the West is really like."[3] He begged off from the job himself for two reasons:

> If I contemplated writing a whole history of the West [his first reason], I would be looking ahead fifteen years. I don't have fifteen years. . . . Such a history would be a major work, and it ought to be done by somebody who could start doing it at the age of twenty instead of coming by it accidentally and sidelong, the way I did. My in-

> volvement in history [his second reason] is personal, not scholarly. . . . All the history and biography that I've done has been an offshoot of personal experiences and personal acquaintances.[4]

And this leads us, finally, to his great work of history, *Beyond the Hundredth Meridian.* "I wouldn't have written Powell if I hadn't known the Southern Utah plateaus," he told Etulain.[5] Nor would he have written it without Francis Parkman's example—Parkman who combined a great narrative style with the personal resolve to walk the actual turf before writing about it, the St. Lawrence Valley for Count Frontenac, the Illinois country for the Sieur de La Salle.[6] In his book Stegner uses the archbooster William Gilpin as a counterpoint to Powell—Gilpin who proclaimed that western wood and water were inexhaustible, that more millions would live in the Great Basin than lived in the Roman Empire at its height, Gilpin "the standard Manifest Destinarian . . . with all his hyperbole, as a sample of the individual who lives by mythology, as against Powell, who was dealing from observed fact."[7]

In *Beyond the Hundredth Meridian* Stegner does identify with John Wesley Powell and his views. Powell affects Stegner's definition of the West and, inevitably, how Stegner wrote history. Houghton Mifflin published *Beyond the Hundredth Meridian* in 1954. Internal evidence suggests that Stegner completed the writing in 1952, when he was forty-three. It is a non-mythological, source-based history in which Stegner explicates Powell's ideas and, without falling into straight-out advocacy, suggests they are relevant to the West of the 1950s. Was he a surrogate for Powell? Thirty years later Etulain asked him: "I gathered . . . that you find his attitudes . . . still very useful?" And Stegner replied: "Oh yes, and still being resisted by a good many forces in this generation. . . . I would say there's a very definite constraint upon western growth, and Powell knew why. Aridity is something that you simply cannot fake out. You can only make maximum use of what water you've got."[8] But because of desertification from overgrazing grasslands, the gross overuse of the Colorado, and other environmental enormities, Stegner concluded in 1982 that "I'm not very optimistic about the future of the West."[9]

Beyond the Hundredth Meridian consists of five chapters in just over four hundred pages of text and notes. Every chapter has narrative drive, the first and longest perhaps the most, because it is chiefly given over to an account of Powell's exploration of the Front Range of the Rockies in 1867–1868 and then his glorious and harrowing expedition down the Colorado in 1869, never knowing what waterfalls, avalanches, or grade-five rapids frothing through vertical cliffs, were around the next bend. Powell did this without the right arm that had been shot off at Shiloh. Stegner, after opening the chapter with a contrast between William Gilpin's egregious boosterism and Powell's realism and science, and after briefly describing Powell's early years in upstate New York, then Ohio, then Wisconsin, then Illinois (a child in a family that wandered much as Stegner's own would do), gets us into the Rockies and the plateau province. We cannot put the book down. We are hooked. We read in the next chapter how surveyors, scientists, artists, and politicians competed for their shares of western action while Powell kept on quietly surveying the plateau province.

Powell regains center-stage in chapter three, developing and presenting his famous 1878 *Report on the Lands of the Arid Region of the United States*, containing his audacious proposals to junk the homestead system and replace it with a combination of four-section rangeland tracts and, in the few places where water existed, small irrigated farms. In chapter four, Powell takes charge of government science, both ethnology and topography, in the early 1880s; in chapter five, which covers the years from 1886 until his resignation in 1894, Powell runs up against special-interest western senators such as William Stewart of Nevada. In Stegner's words, Powell charged into the line like a fullback and gained one yard—and then was thrown for losses. His plan was twisted out of recognizable shape when, shortly before his death in 1902, Congress passed the Newlands Reclamation Act. In the final chapter, called "The Inheritance," Stegner tells what happened and will inevitably happen as the plains erode, as the Colorado silts up behind the big dams, and as the conflict between science and the public interest on one side, and Gilpinesque development and private interests on the other, continues. The book is a history with great present relevance.

It is a fine history not only because it is source-based and Stegner knew the turf. It is a great *story*, as exploration stories almost always are. The space in question, the plateau province, is almost unsurpassed in grandeur. Everyone has seen Monument Valley, knowingly or not, in John Ford movies. Stegner portrays Powell as a peerless leader, sane and scientific yet visionary in his policies, rising to the top of government science amidst lesser men and against bad guys, the narrow, myth-ridden, special interests, who ultimately lay the hero low but cannot slay his enduring ideas. All this time, the writing style delights the reader and renders the prosy historian envious. Allow me to cite just a couple of sentences I wish I had written or had the nerve to write. On the eclipse of Ferdinand Hayden, an earlier expedition leader and darling of Congress:

> Shut out from the inner councils, misinformed by his Washington scouts, Hayden had already been unhorsed, but neither he nor Congress knew it yet.[10]

On Hayden's futile attempts to undermine the Powell Report:

> The tearful defenders of the little man with 160 acres and a plow misconstrued the intention of the proposal completely and either through misunderstanding or malice pictured it as the preamble to landlordism. [p. 259]

And therefore:

> The winds blew through the halls of Congress and the myths were invoked and the shibboleths spoken and the gospels reasserted. [p. 239]

In connection with Powell's ethnographic work in the 1880s, Stegner deftly, searingly, sums up the day's attitudes toward Indians:

> One of the most obvious facts of history to the white Americans who by discovery, exploration, trade, bullets, rum, treaties, and the Word of God took over the continent from its aboriginal inhabitants was that the aboriginal inhabitants were doomed to extinction, and soon. The Kansas editor who prayed for the day when Lo and all his tribe should be obliterated felt that though the day was unwarrantably delayed, yet he could rest in hope. [p. 256]

One could easily multiply examples. If even 10 percent of our own books were so well written, we historians would never again fear competition for enrollments nor lament the job market.

As Stegner pointed out, *Beyond the Hundredth Meridian* is not a conventional biography of Powell but the story of his career and how his ideas developed, especially his ideas about the "arid region." It is not about the Pacific Coast or even the Sierras, but the Colorado basin, an area not urban in the slightest, not a war theater of any note because its Utes and Paiutes were so few, not a place to plow or irrigate. It was, and is, the "deep West," the "unambiguous West."

Would Stegner have been happier if Powell's proposals had been adopted into law—rather than bent out of shape in the Newlands Act of 1902 and those land acts between 1894 and 1916 that tried to split the difference between traditional homesteading and Powell's grazing tracts? Stegner says they were in fact adopted. The Bureau of Reclamation, he says, "is such a bureau as Powell himself might have proposed" (p. 354). But that was written in 1953, and I doubt that a mainstream environmentalist like Stegner would have said the same thing forty years later. In any case, he is a little more guarded at the end of the book: "Powell's program itself has been bent; it has had to swerve and sometimes backtrack; it has succeeded only partially, or in changed proportions; its motion has sometimes whirled around a vortex of failure. But it persists, and moves" (pp. 365–366). Although Stegner closes the book by affirming that Powell was ahead of his time, and we can "with some confidence wait for the future to catch up with him" (p. 367), we know that in 1982, talking with Etulain, he was not so confident about the West.

But would he have been confident if Powell had won decisively, his plan implemented unbent? Perhaps not. Paul W. Gates, the dean of historians of the public domain, once wrote that Powell underestimated the wheat-growing capacity of the area from the 100th meridian to the Rockies. Thus his proposal of four-section grazing ranches would have been subject to widespread fraud because ranchers would have converted them to wheat farms when the price was right.[11] Once plowed up, the grassland would have been hard to restore. (Many grazing ranches were in fact turned into wheat farms at some point. That was one cause of the Dust Bowl. Would any more of the public domain have been misused under Powell's plan? It is not clear how.)

As for his irrigation tracts, Powell never claimed, as Stegner understood well, that more than 2 or 3 percent of the West (however you define it) could ever be irrigated. And most of that land has indeed been irrigated under the Newlands Act and its successors. What might have been different under Powell's plans? Perhaps more federal control for a longer time. But Gates would probably say: no, they would have been transformed too, enlarged or abandoned, as market advantage suggested. Once they were turned over from public to private ownership—without the strong regulation that Americans (above all westerners) always resist—the lands would have been used for private advantage.

By the time he wrote *Where the Bluebird Sings to the Lemonade Springs*, Stegner's optimism had ebbed—if not noticeably since the conversations of 1982, then from *Hundredth Meridian* in 1952. Despite the ever delightful style and generosity of spirit in *Bluebird*, in his dudgeon about misuse of the West and its water, its overcrowding, he loses his firm grasp on realism and succumbs at last to myth. He is outraged about the metropolitan West. But he knew in 1982 that most of the West is metropolitan. It cannot be denied; there is no turning away from it. In the coming century it will become even more urbanized, and its population may reach 94 million by 2050, according to the Census Bureau's mid-range projection.[12] Perhaps we need a new Powell to set forth a dual comprehensive plan—one for the wilderness West and one for the metropolitan West—as the old Powell did for the grazing West and the irrigable West. Even if such a new Powell plan were twisted out of shape, as his was, it would improve on the present absence of one. The magnitude of the task makes even Powell's job, and his vision, appear very simple. But as Stegner realized at least most of the time, both Wests must be recognized and dealt with, the urban as well as the arid. Realism demands no less.

CHAPTER EIGHT

BIOREGIONALIST OF THE HIGH AND DRY: STEGNER AND WESTERN ENVIRONMENTALISM

DAN FLORES

> HOMESICKNESS IS A GREAT TEACHER. It taught me, during an endless rainy fall, that I came from the arid lands, and liked where I came from. I was used to a dry clarity and sharpness in the air. I was used to horizons that either lifted into jagged ranges or rimmed the geometrical circle of the flat world. I was used to seeing a long way. I was used to earth colors—tan, rusty red, toned white—and the endless green of Iowa offended me. I was used to a sun that came up over mountains and went down behind other mountains. I missed the color and smell of sagebrush, and the sight of bare ground.[1]

Americans, a modern de Toqueville might be tempted to think, have a national culture more than we have a geography. Here at the turn of the twenty-first century—more than five centuries after indigenous American Indians and Europeans first eyed one another speculatively on the far island rim of the continent—precious few of us have stayed put long enough to become natives of place. Thirty percent of the American population is in the process of relocating every year. Like migrating Comanches and Kiowas in the nineteenth century—peoples whose sense of topographic significance, power places, and sacred spots existed as an idealized mental map and was thus transferable from one landscape to another—over those five centuries Euro-Americans have done more superimposing than learning and adapting. A Euro-intertwining with the continent, which can be said to have begun with Jefferson on his mountain with the Blue Ridge lining the horizon and continued with Emerson admiring the fact that Thoreau and his brother knew the Merrimack and Concord from end to end, is still nascent in the United States. Movements like bioregionalism and watershed consciousness are less than two decades old. We have far to go in becoming natives.

Wallace Stegner was for most of his career considered a "regional" author, a writer about places whose residents wear Levis rather than tweed and carry Skoal instead of pipe tobacco in their shirt pockets. That's an instantly damning cast of characters to literary critics across much of the country. But Stegner never seemed to blush at it, perhaps because his typically transient twentieth-century American experiences had carved out an enormous, complex, and rich regional homeland for him to write about. It seemed to pain him, as I will point out, that his western roots were mesquite-wide rather than juniper-deep. But as both a practical writer and a literary environmental activist, Stegner drew strength from the fact that he had lived widely across the West—a strength he augmented by careful perusal of both the paper trail and the lives of his region. These traits make him something of an inspiration for the West's upcoming century. He was our model of a proto-bioregional citizen of the larger West.

What he really represented, I think, was a transition figure in the tradition of the great American literary environmentalists of place—a group that commences with Thoreau sitting in his cabin by Walden Pond and includes John Muir on the lip of Yosemite, Aldo Leopold planting pines on his Wisconsin farm, Mary Austin admiring southwestern "village socialism," and Rachel Carson marveling at the sea life along the New England coast. Stegner was a transition figure because he took a larger country—almost (not quite) the whole American West—for his place and because he lived long enough to see environmentalism escape the grasp of the literary amateurs and become something of a modern profession.

But reading Stegner's books, you eventually realize (perhaps not for a long time, so gracefully are the ideas interwoven with the language) that although his experiences were extensive across the West, he was most interested in western places that bore a recognizable stamp based on a definition that existed in Stegner's head. And because that definition existed as a coherent idea in his imagination, it spun out onto the pages he wrote, and into our heads, too.

My own introduction to Stegner's western vision came with a graduate school reading of the most historical work he ever wrote. *Beyond the Hundredth Meridian* is the life story of the nineteenth-century scientist-explorer and federal land manager, John Wesley Powell. But it is far more than that. It is, as well, an adventure story—about an extraordinary part of the world that in those pre-Abbey days was still known as the Colorado Plateau—and a history of ideas that broke Stegner (and open-minded readers) free of provincial innocence about the West. *Beyond the Hundredth Meridian* not only opened a lot of eyes about western realities, it was a work of high art whose flowing, forward-tilt narrative and clear-eyed logic possessed the irresistible pull of the Colorado River itself.

Among the most remarkable things about this new, critical vision of the West was the consistency Stegner gave it and the extent to which (as happened with Aldo Leopold's evolving knowledge of ecology) it became *our* vision. Thirty-five years after the Powell book, after he had already won his Pulitzer for *Angle of Repose* and after he had already trained a legion of fine writers such as Larry McMurtry, Edward Abbey, Scott Momaday, and Wendell Berry, Stegner published a magisterial little essay collection titled *The American West as Living Space*. As I began working on my own response to Stegner shortly after his death, I picked up my heavily marked copy of this slender volume. And at first opening I encountered a paragraph of marginalia written on a blank page at the end of the first essay.

This is how *The American West as Living Space* struck me in 1987: "All the traditional characters are here—Turner, Webb, Powell, Mary Austin. But he also introduces us to William Least Heat Moon, Leslie Silko, Gary Snyder, Bernard Malamud, Don Worster, James Welch, Marc Reisner. His West had a continuity, and what happens next is maybe more important than what's already happened." I went on: "No one has ever synthesized the environmentalist position with respect to western development more thoughtfully, at a more informed level, or with more graceful language." And then I'd written these lines: "Flying out of Chicago, bound for sagebrush and the Llano Estacado and my canyon homestead out on the western plains, I flip through this little book with the same reverence these Southern Baptist conventioneers on the plane around me are lavishing on their New Testaments whilst bound for the mecca of Dallas." I suspect I am like a lot of others in internalizing Stegner's ideas that way. And notice how my language was subtly influenced: western plains; canyon homestead; Llano Estacado. Those words imply a vision I would like to examine here.

The essay that inspired this marginal commentary was titled "Living Dry." In twenty-four pages it distilled and imparted, in those memorable Stegnerian cadences and rhythms, what seems to be an entirely complete vision of the American West and its trajectory through time. Despite its grace of presentation, that vision was tough-minded, steeped in a knowledge of the manifold choices that make up history, and in its sum amounted to a pretty caustic social criticism. It was a vision, with perhaps only one or two exceptions, which most modern western environmentalists would be proud to endorse. Yet Stegner's vision had a power that many environmentalist manifestos lack: it was a product not just of a far deeper knowledge of history but of more penetrating thinking. Not *infallible* thinking, though, as Stegner himself certainly knew.

Stegner's career gave testimony to at least three key insights about life in the American West. All three sprang, I believe, from a confluence of his personal experience and awareness as a westerner, but perhaps even more from his disciplined reading of a core of seminal western thinkers: John Wesley Powell himself, in love with the Colorado Plateau and Mormon adaptations; Frederick Jackson Turner, with his Darwinian ideas about how societies resemble species in their responses to habitats; and the Texan, Walter Prescott Webb, who had argued persuasively in the 1930s that the West was unlike anywhere else on the continent by virtue of its environmental characteristics. It is a tribute both to Stegner's internalizing of these ideas and to the power of his particular expression of them that all of his insights remain at the core of the debate about environmentalism in the West today.

Two of Stegner's principal themes—the West as America's unique public lands province, in which all of us have a stake, and the importance of western wilderness to American character—I would like to visit briefly. The third, which might be called the "Adapting-As-Natives-To-A-West-That-Is-Fragile-By-Virtue-Of-Aridity" insight, and which undergirds the others, requires more examination.

It was not only the act of understanding John Wesley Powell's life that set Wallace Stegner on the course that has influenced western environmentalism ever since. There was also the influence of his friend, fellow Utahan Bernard De Voto, whom Stegner met while teaching back East and who honed Stegner's sensitivity to the role of the public lands in western history. At Wisconsin and Harvard at midcentury, western issues like regionalism and threats to the public lands were nonstarters. But De Voto kept Stegner abreast of what was happening out West, particularly in the late 1940s, when Congressman Frank Barrett of Wyoming was hatching his extraordinary plan to dismantle Grand Teton National Park, then all western national forests and Grazing Service lands, and have the West emulate Texas and privatize everything. From his position as columnist for *Harper's*, De Voto became the first to make defense of the public lands system in the West a foundation of modern environmentalism, fighting the "land grab" as he called it with sarcastic pen. When Stegner wrote him a letter expressing his own outrage, De Voto's response, Stegner recalled, was an immediate reply, "If you feel like that, goddammit, get into print with it—we need everybody in print that we can get." And he did, because "my guru had spoken."[2] Stegner's career as an environmentalist had begun.

Hence, the influences: brought up in a variety of western places, being exposed to high literature and culture in the East, reading Walter Prescott Webb, researching Powell's life, harkening to De Voto's call over threats to western public lands. They all coalesced in a mature, reasoned, quiet but in retrospect undeniably passionate response when in the early 1950s dams threatened a part of the public lands that Stegner, among few others, knew intimately. In 1946, the year Stegner had effused in print over Hoover Dam, the Bureau of Reclamation published a report titled "A Natural Menace Becomes a Natural Resource."[3] The menace in question was the Colorado River, and the people who fantasized about its conversion to natural resource were the citizens of Colorado, Utah, and Wyoming, who believed that the federal government ought to build them dams as it had done in Nevada and California.

The Democratic administration of Harry Truman had envisioned the Colorado River Storage Project (CRSP) on the upper Colorado as a TVA for the West, though that vision would change as the 1950s dawned. Economic promises aside, one of the things the CRSP dams certainly were going to do—and only thirty-five years after John Muir's famous battle to keep Yosemite National Park's Hetch Hetchy Canyon from becoming a reservoir—was to back waters up into Dinosaur National Monument. The battle over this issue raged for five years, in the process creating much of what we now call modern environmentalism. The Dinosaur National Monument battle produced conservationists who were dismayed to learn, as Muir had been, what an environmental threat federal development policies could be. It taught environmentalists that engineers and their science were not infallible. And in those pre-NEPA, pre-EIS, pre-Endangered

Species Act days, it reaffirmed that poetry, prose, and pictures were often the best way to swing a battle in which the public had a stake.

Looking back now, it almost seems as if fate had poised Stegner at this juncture in his career to assume De Voto's mantle as America's environmental poet laureate. De Voto launched the literary crusade against incursions into Dinosaur National Monument with his "Shall We Let Them Ruin Our National Parks?" which appeared in the *Saturday Evening Post* in July 1950. But as the Sierra Club and its director, David Brower, were to learn, only a handful of people really knew this Colorado Plateau country well—and one of them was Wallace Stegner, whose biography of John Wesley Powell was then under way (it would appear in 1954) and who in response to De Voto's call had already contributed articles to *The New Republic* and *The Reporter* condemning the project. Stegner had retraced much of Powell's route down the Colorado and had otherwise immersed himself in the topography and natural history of the wild slickrock country the CRSP threatened to transform. So as Brower faced the prospect of galvanizing American conservationists around a landscape of desert and naked rock—conservationists for whom wilderness then was largely synonymous with *mountains*—he hit on the idea of a coffee-table book coupling fine writing and stunning photography to show Americans what was at risk. To introduce, edit, and help assemble the essays of *This Is Dinosaur: Echo Park and Its Magic Rivers*, the first in what would be a long line of such books produced by the Sierra Club, he turned to Wallace Stegner. Alfred A. Knopf got the book out in 1955 and sent every member of Congress a copy. By the time the CRSP bill was signed in 1956, seventy environmental organizations had assembled against the two dams. Muir had mustered support from just six groups against the Hetch Hetchy dam.

In the final version of the plan, of course, the offending dams were dropped and Dinosaur was saved, but the peerless Glen Canyon had to be sacrificed—or so environmentalist mythology goes. In fact, a dam at Glen Canyon was always part of the CRSP; the only thing environmentalists really bargained away was some additional height in the Glen Canyon structure. Nonetheless, as one of no more than two hundred people who were familiar with Glen Canyon, Stegner always regarded it as a personal failing that he had not fought harder to keep that quiet sandstone marvel from being inundated. Stegner's role in these events went far toward making him into a younger version of the literary environmental guru De Voto had been. As a result of his work on *This Is Dinosaur,* Stegner was made an honorary life member of the Sierra Club and was even elected to its board of directors. Brower involved him in the Exhibit Series of books the Sierra Club began to publish in the 1960s. And in 1982 he was awarded the prestigious John Muir Award from the country's best-known environmental organization.

By his own assessment, it was because he knew western history as he did that Stegner became famous as a literary champion of still another idea that continues to define western environmentalism. In the late 1950s, after three decades of agitation by such conservation figures as Aldo Leopold, Bob Marshall, and Arthur Carhart, Congress began to hear testimony on bills to create a wilderness system in the United States. Stegner was by no means the only voice, or even the most important voice, to speak about the necessity of wilderness to American culture. But his 1960 "Wilderness Letter," now a worldwide classic of environmental history, was a galvanizing document and with good reason has been endlessly quoted as perhaps the single best statement on behalf of wilderness preservation. "I want to speak for the wilderness idea as something that has helped form our character and that has certainly shaped our history as a people," Stegner wrote.[4]

In one sense Stegner's was largely an anthropocentric argument: preserving wilderness was crucial for Americans, he believed, not so much because it set aside habitat or preserved slices of the primeval earth, but because American history has placed the idea of wilderness so close to the spiritual and mystical core of who Americans are. In the process of conquering the continent we surrendered our souls to the wild and the natural—and this has given Americans a sense of bigness, a primal sense of ourselves as "a wild species" that few modern peoples retain. Preserving wilderness amounted to a "geography of hope," as Stegner's memorable phrase put it, a hope that humans could retain that sanity of knowing they are nature's children. Echoing Thoreau, Stegner believed that the wild held out to Americans our last, best hope of being "good animals."

In 1961, after a decade of Republican presidents and development-minded secretaries heading the Interior Department, Stewart Udall took over Interior in the Kennedy administration. Stegner was encouraged enough to send him a copy of *Beyond the Hundredth Meridian*—a reading of which, it is rumored, has now become something of a tradition for new stewards at Interior. Udall was impressed enough to persuade Stegner to come to Washington to serve, briefly, as a special assistant. While there Stegner helped Udall research and plan a book for which Udall has been famous ever since. *The Quiet Crisis* appeared in 1963. Along with Rachel Carson's *Silent Spring*, it is often singled out as playing a major role in the phenomenal surge of environmentalism as a widespread sociopolitical issue in the modern United States. But Stegner did more than that at Interior in the first years of the 1960s. And in at least one of his other tasks one can sense the division of the generations that would separate his legacy in environmental writing from that of, say, an Edward Abbey.

In the years when Abbey was serving as ranger at Arches National Monument and preparing to write a book (*Desert Solitaire*, published in 1967) attacking plans to create for "industrial tourism" more national parks in the Utah canyonlands, Stegner and Joe Carrithers were, at Udall's direction, making the very tour of the canyonlands that resulted in recommendations to expand and upgrade monuments like Arches and Capitol Reef to national park status. Creating a national park around Capitol Reef, in fact, was one of Stegner's fondest projects—a chance, he probably thought, to play John Muir, but in *his* country, a place he knew as no one else did. Perhaps, as Abbey insinuated, sentiments like these were naive and misplaced. But Stegner did not change his mind about the role that national parks could play for America. Only a few years later he visited Escalante Canyon and wrote again of his hopes for creating a national park in the higher elevations of the country between Capitol Reef and the northwestern shores of Lake Powell.[5]

Lacing together all these issues involving parks, wilderness, and public lands in the West was Stegner's personal fascination with aridity. Among the most simple yet compelling ideas that motivated Wallace Stegner seems to have been his conviction that the West is, environmentally, a very different place from the rest of the United States—and that aridity is the cause of that difference. "Aridity, and aridity alone, makes the various Wests one," he wrote.[6] It is aridity that accomplishes the seemingly magical transformation of the landscape beyond the 100th meridian, aridity that has served as the principal challenge to living in the West, aridity that in Stegner's view imparted to the modern West its defining identity as the domain of the public lands. The fact of dry across much of the West convinced Stegner that the West was, in truth, far more a fixed place on the map than it was a frontier process working its way across the continent, à la Frederick Jackson Turner's famous frontier hypothesis.[7]

For the aesthetics of landscape, the twentieth century eventually will be recognized as the century when Americans discovered their deserts. Anyone who reads nineteenth-century exploration and travel accounts knows the disgust and alienation in which most westering Americans held the High Plains and interior deserts of the West. The reasons were not strictly utilitarian. American landscape sensibilities in the nineteenth century were shaped by the worship of forested mountains in the tradition of John Ruskin and the great mountain painters of Europe like F. W. M. Turner, Claude Lorraine, and Salvator Rosa. That aesthetic had been confirmed in America with the popularity of the Hudson River school of painters. When pioneering midcentury landscapists such as the painters Albert Bierstadt and Thomas Moran and the photographer William H. Jackson went West, it was no accident that the body of work they executed came to be known as the Rocky Mountain school. The arid plains and deserts did not draw them. As for homesteaders and settlers, their myths ("Rain Follows the Plow," "Making the Desert Bloom As the Rose") and institutions (William Smythe's Irrigation Congresses) were all about denying the desert or transforming it. An appreciation of aridity aesthetics was to take another century. It would find expression in desert lovers like Mary Austin (*The Land of Little Rain*, 1903), John C. Van Dyke (*The Desert*, 1905), the work of the modernist painters of the Santa Fe and Taos art colonies, and particularly Georgia O'Keeffe's emotionally charged landscapes of West Texas and New Mexico badlands country. Several of Joseph Wood Krutch's books on the western deserts had appeared in print, too, by the time Wallace Stegner began his work in the 1940s.

As a native Louisianan whose childhood surroundings ought to have engendered a particular horror for arid country, I know that in the development of my own landscape aesthetic—and I suspect that I am not alone—Stegner's was a crucial voice. My choice of a canyon homestead in the Southwest in the early 1980s owed much to Stegner's influence. When I moved to the Montana Rockies and began looking for land a decade later, I instinctively sought out the most arid setting within commuting distance of the University of Montana. I write these lines today in the eastern foothills of the Bitterroot Valley, surrounded by sagebrush and rabbitbrush, ponderosa pines and junipers, partly because of Wallace Stegner. He loved what lack of moisture did to topography, air, light, life. The West has desert at its heart. And Stegner seemed so attuned to the principle that his books, at first opening, always seemed to carry just a hint about them of dry wind and the nasal burn of sagebrush crushed between the fingers. That aesthetic is my own, and it sends me and legions of others to make annual pilgrimages to places like the Utah canyonlands or the badlands of eastern Montana, where dry draws its influence across the West in powerful and soul-stirring ways.

Graduate training in Texas has given me hints as to where Stegner absorbed many of his own ideas about aridity. Aside from the works of Austin and Krutch, the text that most vividly set forth the aridity theme during Stegner's formative years was Texas historian Walter Prescott Webb's 1931 book, *The Great Plains: A Study of Institutions and Environment.* But unlike Webb, who in the tradition of the nineteenth century presented the West's aridity as a negative, I read in Stegner an admiration of aridity, an assessment of dry as a positive good for the West in that it imposed limits. Aridity has forced westerners—as no other Americans have been forced in quite the same way—to address limits, to acknowledge that the world *is* (so to speak) flat and has borders. To the end of his career Stegner fought for accepting rather than trying to modify the limits implied by aridity.

Stegner wrote about aridity not only well but often. This passage is from "Thoughts in a Dry Land," in *Where the Bluebird Sings.*

> You know also that the western landscape is more than topography and landforms, dirt and rock. It is, most fundamentally, climate—climate which expresses itself not only as landforms but as atmosphere, flora, fauna. And here, despite all the local variety, there is a large, abiding simplicity. Not all the West is arid, yet except at its Pacific edge, aridity surrounds and encompasses it. . . . Aridity, more than anything else, gives the western landscape its character. It is aridity that gives the air its special dry clarity; aridity that puts brilliance in the light and polishes and enlarges the stars; aridity that leads the grasses to evolve as bunches rather than as turf; aridity that exposes the pigmentation of the raw earth and limits, almost eliminates, the color of chlorophyll; aridity that erodes the earth in cliffs and badlands rather than in softened and vegetated slopes. . . . The West, Walter Webb said, is a "semi-desert with a desert heart." If I prefer to think of it as two long chains of mountain ranges with deserts or semi-deserts in their rain shadow, that is not to deny that the primary unity of the West is a shortage of water.[8]

Of course the American West is not a uniformly arid province at all, and this fact complicates macro-interpretations of the West that rest on environment alone. The West is a wild diversity of places, most arid, many not, including many settled high-elevation parts of the Mountain West where water and green almost replicate eastern conditions. It does possess deserts, and slickrock canyons, and much vast and open shrubland of greasewood and sagebrush. And the Great Plains, at least west of about the 98th meridian, in places still present badlands and great grassy sweeps rivaled only by trans-Ural Asia as the premier dry grasslands of the world. But not all of the West is desert-influenced. It consists, as Stegner did note, of a Pacific Northwest that in some respects replicates the environmental conditions of Western Europe. The Southern Cascades, for example, produce more annual runoff than most of New England. And as Stegner knew, the Interior West is many Wests—up to nine ecological life zones, zebra-striping rows of serrated, oval, table-topped mountains rising like wet emerald islands from the lowland sea of plains, and sagebrush steppe, and cactus desert.[9]

Stegner recognized this diversity but clung primarily to the aridity theme. That elemental premise—that it was not any lack of land, the so-called Closed Space argument, but a scarcity of water that defined western history and possibilities—led Stegner to anticipate the kinds of ideas environmental historians have made their careers analyzing. To his credit, Stegner came to realize what all this meant in terms of the larger environmental picture for the West, for adaptation and carrying capacity in particular.

Stegner did not emerge early in his career a full-blown western social critic, ridiculing the big technological fixes for aridity—the dams that have engineered so many western rivers out of existence. In fact, in 1946, when he was commencing his work on the Powell book, he was impressed by his visit to Hoover Dam, writing in *The Atlantic Monthly* how that megalithic expanse of concrete had given him a "World's Fair" feeling. While Powell had envisioned some one hundred and twenty hydrographic commonwealths in the West, a positive vision some would say, he had imagined them held together by water manipulation and hydraulic engineering. But Stegner was learning to be more critical. Eventually he would refer to dam building in the West as "original sin" and emerge as an acute critic of the grandiose plans for diversion projects that promised more water from the Southeast or the Pacific Northwest or even the Yukon in order to transform the West into something it was not. Thus his work in protecting Dinosaur National Monument rested logically on his growing conclusion that the West needed to live with and embrace aridity.

Stegner's vision of how life ought to be lived in the West, derived from his Webb–Powell–De Voto influences, evolved over time, reflecting his experiences. But the foundations of the ideas were always consistent. He had always been fascinated by place in the West and by the complex interaction of human cultures with the variety of environments that made up the West. In some of his last nonfiction—*The American West as Living Space* and *Where the Bluebird Sings*—he urged a kind of organicist adaptation for western societies. From his consistent idea that aridity was at the core of the West came the idea that because of aridity the West is substantially unique. And it follows that aridity makes the West more fragile, that its resources must be managed more carefully, that it cannot absorb the levels of environmental modification or human population the rest of the country can.

It is impossible to read Stegner without concluding that most of his big ideas sprang as much from the observations of a soul that watched the world ceaselessly as from reading books. So we should not be surprised that his championing of the next logical step in western adaptation, the process of going native that he liked to call developing "a sense of place," should have begun for him on the Saskatchewan prairie where he spent his childhood years. All the sensuousness he associated with humans who identify themselves in the context of a piece of known earth—the haunting smell of wolf willow, for example—had its origins there. "I seem to have been born with an overweening sense of place, an almost pathological sensitivity to the colors, smells, light, and land and lifeforms of the segments of earth on which I've lived," he wrote.[10]

Listening to his audiocassette *A Sense of Place*, I am struck with two impressions. First, Stegner knew firsthand the dangers inherent in the transience that is such a part of the American character. How can Americans ever expect to go native, he asked, when we never sink roots anywhere? "Our migratoriness has robbed us of the gods who make places holy," he once wrote. As for his own transience, he regretted it, thought it had value only in the sense that it had perhaps made him a native of the whole West—and still, late in life, decided to pick one of the many places where he'd lived, Salt Lake City, and claim it for his own.

I suspect that this choice involved more than an old man's nostalgia for his adolescent memories. Another impression emerges from his books: he admired western societies that had created unique responses to their local environments. Missoula, Montana, with its piquant mix of writers and environmentalists, is one such place that he singled out in this regard. But it was Mormon Utah—a very different West from progressive Missoula—that focused Stegner on the dialogue between place and culture. As he explained to Richard Etulain in *Conversations with Wallace Stegner*, it was not until he left Utah that he realized what a unique lifeway had resulted from the interplay of Mormon culture and the ranges and canyons of the Intermountain West. *Mormon Country* was his introductory exploration of that phenomenon, which deepened considerably in *Beyond the Hundredth Meridian*, when he realized that John Wesley Powell's blueprint for a kind of planned, Darwinian adaptation to the West was based largely on Powell's admiration for the Mormon model.

Stegner had learned to be suspicious of the shibboleth of western individualism. And here, in the communally based Mormon towns, he found the growing seed of one distinctive sense of place in the West. This recognition led him to look elsewhere for the phenomenon in other cultures he believed were in the process of going native in the West—the New Mexicans of the mountain villages around Taos and Santa Fe, the ranchers of the Absaroka country in Montana, small western cities with flavorful personalities like Missoula and Corvallis, Oregon. He did not always applaud the details of the process, as in frantically booming modern Utah. And it distressed him (as he told T. H. Watkins) that it rarely seemed to be the longtime natives of such places that stood up for their environmental protection.[11] But the process of creating places out of spaces always interested him, so that even *American Places*, coauthored with his son Page in 1981, did not stand as his last word on the topic.

In fact, in his dissonance over whether he had ever managed to go native himself, he wrote a superb essay about creating a small ranch in the Los Altos Hills outside San Francisco that can serve as a guide to a conscious immersion in place—or what today we would call western bioregionalism. There is a bit of irony in the circumstance. Many environmentalists in the 1990s regard the creation of rural homes in subdivided natural areas as one of the greatest threats the West currently faces, and yet in his own effort to go native Stegner himself could not resist doing that very thing. In a reprise of the homesteader experience, however, and like so many of us across the West today, Stegner in essence hoped to be the last homebuilder allowed in his area. There was a great healing in that early absence of lights and presence of coyotes and hawks and deer around his place in the Los Altos Hills. But as the houses and vapor lamps inevitably drew closer, he helped found the Committee for Green Foothills to slow their approach.

More than his awareness of the limits imposed by drylands living, or his hope for a bioregional sense of place in the West, was Stegner's appreciation of wild places and their effects on humans that made him a kind of model westerner. You have to suspect that what Stegner most wanted was to be left alone to read and experience and write. But the circumstances of his time and his place made it impossible for him to hide in the scholar-writer's life. If he was, as many environmentalists knew him to be, a reluctant activist, I think we ought to note how often he set his reluctance aside and inspired western thinkers and writers of our time. Encouraged first by De Voto then perhaps by his own sense of morality and mission, Stegner never stopped the flow of environmental essays, usually about wilderness and public lands issues, that came from his pen. And somehow, too, he found time to serve not just on the Sierra Club board of directors, and as a special consultant to Stewart Udall, but on the governing council of the Wilderness Society, on the councils of the Trust for Public Land and the Greenbelt Society, and on the Advisory Board for National Parks, Historical Sites, Buildings, and Monuments.

It is a simple thing to say that Stegner's influence was profound. And like the big, elemental ideas he had about the West, there is an irresistible force in simplicity. But as Edward Abbey once observed, writing books and articles isn't enough. Knowledge isn't enough. Like Bernard De Voto for an earlier generation, like Abbey for a somewhat different cultural subset today, Stegner has served as an apotheosis of environmental activism for writers and even academics whose topic is the West. He married unusual intellectual rigor and literary grace with social commitment, and in his quiet way he promoted those—his friends K. Ross Toole and William Kittredge in my part of the West—who shared that commitment.

There are those in the West who continue to scoff at ideas involving adaptation, wilderness, and bioregionalism. If they are now confronted by a core group of talented western writers that has coalesced in support of these very issues, blame Stegner. As Stewart Udall recently said to me about Stegner's legacy, Wally was a good animal, and that made him optimistic about the West despite his fears for the region. Stegner thought that if anybody had a chance to save the West, it was all those minor regional writers in love with all those western places—those legions of us who have been inspired and strengthened by the Grand Old Man of western environmentalism.

CHAPTER NINE

WALLACE STEGNER: GEOBIOGRAPHER

CURT MEINE

AT THE END OF HIS *Conversations with Wallace Stegner*, Richard Etulain posed the inescapable question: "What has Wallace Stegner attempted to leave for his readers?" Stegner must have seen it coming. He responded bemusedly to Etulain's effort to dredge up "the philosophical residue, the sludge at the bottom of the cup." Stegner then offered up this concise summation: in his writing he attempted "to say how it was," to represent "the human response to a set of environmental and temporal conditions."[1]

Stegner examined that human response through all the genres in which he worked—novels, essays, short stories, criticism, memoir, history. One senses that, like a batter who ignores the bunt sign and hits a home run, Stegner had a mind of his own when it came to literary strategy, and would not have confined himself to any one of these approaches even if his editors, critics, and readers had told him to. The techniques available through these varied genres allowed him to respond to opportunities with all the skills he possessed, to meld his voices to suit the story he wanted to tell, the truth he felt compelled to explore.

Stegner also seized the opportunity to examine the "human response" through his biographies. In a sense, biography may have been the form for which the genre-bending Stegner was best suited. "Truth, Candor, and

Honesty," Virginia Woolf once wrote, are "the austere Gods who keep watch and ward by the inkpot of biography."[2] In his definitive commentary *Writing Lives*, Leon Edel distilled into a single sentence the creative tension that drives biography and with which every biographer must learn to live: "A writer of lives is allowed the imagination of form but not of fact." In this sense, Edel writes, "the writing of lives is a department of history and is closely related to the discoveries of history. It can claim the same skills. No lives are lived outside of history or society; they take place in human time. No biography is complete unless it reveals the individual within history, within an ethos and social complex."[3] But the skills of the novelist also come into play as the biographer seeks to illustrate character, motivation, conflict, growth. Achieving at once the veracity of history and the narrative strength of fiction is the biographer's challenge.

Compounding the challenge is the fact that each biographer's task is unique and requires an approach particular to the subject. Each subject calls forth a new species of biography fitted to the life, times, and historiography upon which it rests. "The biographer truly succeeds," Edel writes, "if a distinct literary form can be found for the particular life."[4] To realize this distinct form, the biographer is allowed—indeed, required—to range as widely as the subject demands. For this reason, biography will always remain a bastion of interdisciplinary endeavor, where integration of knowledge is of the essence.

Now we can begin to comprehend the task that Wallace Stegner took on when he chose to write the lives he did. For his two major subjects, John Wesley Powell and Bernard De Voto, it was not enough to reveal them only "within history, within an ethos and social complex." In the sub-sub-genre of environmental biography (if such a thing can be said to exist), proportionately greater attention must be given to the spatial dimension, to the environmental conditions of the life under examination. For Stegner, the writing of lives was as much a department of cultural geography and natural history as it was history per se. Slightly amending Edel, we might say that no lives are lived outside a biophysical and biogeographic setting; they take place within a geological and ecological, as well as social, complex. The ethos we must take into account pertains not only to the human community but to the larger and longer-lived community of life in which human lives and cultures are embedded.

These peculiar demands stand in interesting contrast to the dominant trends in the development of biography in the twentieth century. One of the tenets of Edel's "New Biography" is that the emergence of psychology has fundamentally altered the biographer's trade, augmenting the toolbox of techniques, redefining both the methods and the product. Edel writes that "biographies which do not use [psychological] knowledge must henceforth be reckoned as incomplete; they belong to a time when lives were entirely 'exterior' and neglected the reflective and inner side of human beings."[5] On this foundation, Erik Erikson and others have built the entire specialty of psychobiography. But all thoughtful biographies now partake, to one degree or another, of psychology's insights into the development and expression of personality.

Yet our lives are never entirely "interior" or "exterior" but always a dynamic interpenetration of both. And while biography has moved toward deeper examination of the inner self, little attention has been given to the environmental context of biographical subjects—the places a subject both shapes and is shaped by. Let us grant that, for many subjects of biography (especially the supermarket best-sellers), it is hard even to think of them as inhabiting a real, vital landscape. Many of the lives we deem worthy of biography seem almost to be defined by the degree to which they are removed from the soil, the waters, the plants and animals. The critical scenes of these lives take place in offices, studios, machines, crowds, arenas, convention halls, and boudoirs. Yet even as psychology has revolutionized our perception of the inner world, advances in the natural sciences have revolutionized our perception of the world around us. Biographies—of some subjects at least—can never be the same.

To a notable degree Stegner wrote his two biographies, not exactly *against* the modern grain of psychologically informed biography, but *across* the grain. Stegner endeavored to connect inner and outer worlds, to limn lives through integration of the personal, social, political, biological, and environmental. *Beyond the Hundredth Meridian* and *The Uneasy Chair* are distinguished not only by their examination of the relationship between biographical subject and biogeographical space but by the way Stegner uses this relationship to examine forces, tensions, patterns, and themes at the heart of North America's cultural development. Place becomes not simply a background against which human lives are played out but a milieu within which, and with which, human beings interact. For most biographical subjects, perhaps, the impact of climate, geology, biogeography, ecology, and anthropology might be considered of little direct relevance. For John Wesley Powell and Bernard De Voto, these issues are central.

HAVING TAKEN ON THE biographical subjects he did, how did Stegner respond to this expanded spatial dimension of biography? Leon Edel again gives us a useful lead. At one point Edel defines biography as "truth seen . . . through a certain temperament. Biography is most itself . . . when it is a life offered us through a personal vision."[6] And so, we must ask another question: what vision, what perspective, did Stegner bring to his studies of these two lives?

The answer involves the large theme that, like a great river, unified so much of Stegner's literary landscape: the attitudes and adjustments of people to the North American land, and the response of that land to those who have lived upon it. But the answer also involves Stegner's conviction that much of any narrative's strength draws upon the interactions between people and their local place. In *One Way to Spell Man*, Stegner writes:

> Identity, the truest sense of self and tribe, the deepest loyalty to place and way of life, is inescapably local, and it is my faith that all the most serious art and literature come out of that seedbed, even though the writer's experience goes far beyond it. Much of the felt life and observed character and place that give a novel body and authenticity . . . comes ultimately from the shared experience of a community or region.[7]

The same, he might have added, pertains to biography. It too gains "body and authenticity" from the power of the subject's life to reflect and illuminate the shared experience of a locality or region.

And so we find Stegner constantly working through the concept of regionalism, which he distinguished from mere provincialism: "You can get stuck in the provincial," he told Etulain. "It's a very small hole. On the other hand if you use the regional as a springboard or a launching pad instead of a prison, then you can be interested in the world; but you're interested in it from, somehow, the Western view, which is different from the East Coast view. The difference is probably instructive. It should be."[8]

This sharp distinction between regionalism and provincialism is recurrent in Stegner's writing and, as we shall see, woven through his biographies. It is what Stegner had in mind when he described William Stafford as "a poet who is clearly western without being limited by his Westernness."[9] It defines the difference between those who strive toward a continental vision and those whom he described as "locally patriotic." It suggests that to develop an appreciation of your region, you must come to know it from both the inside out and the outside in. "The only way to get perspective on your own culture," Stegner observed, "is to step outside of it." In the case of the American West, he added, "you're stepping out of your own culture and into some culture which is vaguely a world culture, or a national culture, and learning to look upon your narrower provincial culture with some kind of perspective."[10] To this breed of true regionalists, Stegner implies, falls the hard work of integrating local interests with broader regional, national, and global interests.

But another issue complicates the Stegner view. Acting as a thematic complement to the sense of place in much of Stegner's work is the sense of movement—felt most dramatically in *The Big Rock Candy Mountain* and *Angle of Repose*. Rootlessness and restlessness are deeply American traits and formative features in Stegner's own early life. He appreciated the difference between colonizing and peopling, between booming and sticking, between seizing "the main chance" and accepting responsibility. He had an unerring feel for the specific pressure points and personalities through which these tensions have been expressed in American history.

The personal vision that Stegner brought to his biographies incorporated all of these enriching and complicating factors: a critical understanding of regionalism; an awareness of the stultifying effects of provincialism; an appreciation of the regional perspective in developing a viable literary tradition; the experience of movement; a sensitivity to issues of personal commitment and responsibility; the connections between all these elements and the conditions (inherent and invented) in the North American landscape. All of these he brought to the biographies. Throughout their lives, Powell and De Voto addressed the primal relationship between American people and the landscape itself. So, too, of course, did Stegner. And this shared experience allowed Stegner to pour himself into their stories with the full energy that biography demands.

WALLACE STEGNER WROTE two biographies. *Beyond the Hundredth Meridian: John Wesley Powell and the Second Opening of the West* appeared in 1954. Twenty years later he published *The Uneasy Chair: A Biography of Bernard*

De Voto. His interest in the possibilities of biography, however, seems to have been present long before. For his dissertation at the University of Iowa, Stegner examined the work of Clarence Dutton, whom he described as John Wesley Powell's "left hand." This early interest brought Powell into his awareness, initiating the process through which Powell would eventually come to occupy center stage. When *Beyond the Hundredth Meridian* finally did appear, it included a biosketch of Dutton inlaid within the book as a whole.

Stegner's next significant brush with biography was his fictionalized account of the final years of labor martyr Joe Hill, *The Preacher and the Slave*. First published in 1950, the book was reprinted in 1959 as *Joe Hill: A Biographical Novel*. To gain a sense of the peculiar nature of *Joe Hill*, we can turn to a bit of bibliographic arcana. Every book entering the Library of Congress carries on its copyright page what is known in the trade as Cataloging-in-Publication (or CIP) information. The CIP data correspond to the Library of Congress' main catalog and allow librarians to place the books on their shelves quickly and properly. Of course, more than a few of Wallace Stegner's books defy the best efforts of librarians and other catalogicians. Open to the copyright page of *Joe Hill* and you will find the following CIP information: its first listing is "Hill, Joe—1879–1915—Fiction"; its second is "Industrial Workers of the World—History—Fiction." As an expression of the tension in Stegner's work as historian and fiction writer, it doesn't get much more austere than this.

And what of biography, the odd orphan of the middle ground? It exists, as Edel adamantly notes, as a distinct genre of its own with its own inherent tensions, forms, obligations, aims, pitfalls, opportunities. *Joe Hill* may be read as an indication that Stegner had yet to arrive at the optimal means of accommodating his varied interests and his possible voices. *Joe Hill* is, by Stegner's own account, a work of fiction. Yet it is one in which its author went to extraordinary lengths in serving those "austere gods" of Truth, Candor, and Honesty—to the point, for example, of being blindfolded and led through a mock execution in Utah's state penitentiary. Further, Stegner separately published two articles detailing the findings from his background historical research. "I knew as much about Joe Hill as I could find out," he told Richard Etulain. "*If I had been writing his biography* I couldn't have gone any deeper."[11]

Although *Joe Hill* is as firmly affixed in place as any of Stegner's fictions, it reveals no special focus on the West or Utah as regions nor interest in the differences between them and other places. Yet the impacts of geography can be found in its genesis and in its impact on Stegner's subsequent work. He conceived the project during his Cambridge years, when he could look back to Utah with both distance and familiarity: "The place of Joe Hill's end was all familiar territory, and the more I read about [him], the more it seemed that I had some kind of natural interest in him."[12]

When the book came out, it was not what most readers expected. It attempted, not to report or explain or vindicate the life of its subject, but to dramatize it. As Robert Keller notes, the book was a fictional treatment of "a real person whom [Stegner] had made violent and immoral, a real person with a real name, a hero who meant much to other people."[13] *Joe Hill* received, in Stegner's words, "a feeble press and no notice and didn't sell anything, and nobody understood it. I was irritated that reviewers thought I was writing a proletarian novel, which I wasn't. They thought of the book as a belated trailer from the thirties."[14] Between these lines, we can read the impact of Stegner's own experience in Madison and the sophisticated yet independent political outlook he gained there and explored in his 1941 novel *Fire and Ice*.

Joe Hill may have been less significant for what it taught Stegner to do with biography than for what it taught him not to do. The lackluster public response and the misreading of critics led him to conclude that he was "on the wrong track" in his fictions.[15] He seems also at this time to have recalculated his bearings on the middle ground between history and literature. At any rate, in his next venture into the territory of biography he stayed well on the documentary side of the border.

THIS RECALIBRATION may have come out of the changed circumstances in the lives of the Stegners. The emerging emphasis on regionalism in Stegner's work seems to have been connected to his own return, in 1945, to the land beyond the 100th meridian. Several of the essays in this volume take their lead from the passage in *The Big Rock Candy Mountain* describing Bruce Mason's cross-country drive west "beyond the Dakotas toward home."[16] Just a few years after Stegner wrote the scene, he and his family lived it out again. Stegner's recollection of this personal migration is itself a gloss on regionalism and worth quoting in full:

> Teaching at Harvard, which should have gratified my highest ambition, didn't fully satisfy because I didn't much like the place where Harvard was situated. I took the first opportunity that offered a

> chance to get back west. . . . I grew up western, and the very first time I moved out of the West I realized what it meant to me. . . . Wider worlds, but with one foot always kept in the center circle. . . . By the time we arrived in Palo Alto I was already involved in the biography of John Wesley Powell, the quintessential student of the West, and had pretty well committed myself to a lifetime of writing about the country I had grown up in. . . . I was at home, where I belonged, and thought I had lived away from home long enough to know where it was and to have some perspective on it. It is not an unusual life-curve for Westerners—to live in and be shaped by the bigness, sparseness, space, clarity, and hopefulness of the West, to go away for study and enlargement and the perspective that distance and dissatisfaction can give, and then to return to what pleases the sight and enlists the loyalty and demands the commitment.[17]

Stegner's return west to take up his post at Stanford, his deepening relationship with Bernard De Voto, and his entry into western land policy disputes in the late 1940s all contributed to the opportunity that Stegner recognized and seized in *Beyond the Hundredth Meridian*: the role of place in telling lives.

The CIP information on *Beyond the Hundredth Meridian* advises librarians first to shelve it under "Powell, John Wesley, 1834–1902," then "West (U.S.)—History—1848–1950," then "Scientists—United States—Biography." In Leon Edel's typology, it is an unusually expansive example of a "chronicle life," the more or less traditional approach to biography in which the biographer integrates documentary materials (usually in chronological order), supplies extensive background materials, and allows the subject's voice to be heard consistently throughout. "The result," Edel writes, "is a work of history created around a central figure . . . [and] a heavy autobiographical component is introduced into the biographical creation."[18]

Stegner's approach in *Beyond the Hundredth Meridian* reflects the peculiar opportunities and limitations that Powell as subject presented. Stegner was motivated by his historical interest in the subject, his familiarity with the plateaulands of Utah, and his mentorship with De Voto. In Powell he found the one compelling figure that could not only bring unity to large and diverse themes in America's cultural development, but also hold the biographer's strong personal commitment. And that it would certainly take. Stegner devoted twelve years to the effort.

Stegner, of course, did not personally know his subject in this case. For a biographer, direct experience with one's subject always involves trade-offs. What one gains in intimacy, one risks losing in detachment. As Edel notes, every biographer must seek to be a "participant-observer" in the subject's life; in *Beyond the Hundredth Meridian*, circumstances dictated that Stegner must be more observer than participant. There were few opportunities to peer into Powell's inner life—and in any case that was not Stegner's primary goal to begin with. The Powell we encounter is the public Powell. Stegner recognized this: "I wasn't really writing a biography of Powell in the sense of personality, I was writing a career, and the career dealt with the plateau province."[19] Where he could, Stegner the frustrated novelist did take full advantage of the dramatic potential of Powell's story. His account of the initial trip down the Colorado is told with a narrative intensity that ranks with that of his account of *Wolf Willow*'s beleaguered cowhands in the story "Genesis." Edel proposes that biographers "borrow some of the techniques of fiction without lapsing into fiction."[20] Stegner would have appreciated that proposition, the opportunities it allows, and the tension it implies.

Taken as a whole, *Beyond the Hundredth Meridian* was unlike any other biography in the annals of American literature. One could find others set in the same region and period. One could even find prominent contemporary biographies of conservationists: Linnie Marsh Wolfe had received the Pulitzer Prize in 1946 for her treatment of John Muir's life, *Son of the Wilderness*. Stegner provided something quite different: a painstakingly reconstructed account that rescued Powell from obscurity and placed him at the very center of the nation's effort to catch up with its own "forward-leaning disequilibrium of advance, progress, boom, growth."[21]

Even as Stegner's own "life-curve" was turning west and rechanneling his intellectual energies, he found in John Wesley Powell's life-curve a trajectory that illustrated the shifting cultural energies on the American landscape. Though born in 1834 in New York, Powell was a son of the Midwest, moving peripatetically among Wisconsin, Illinois, and Ohio until the Civil War, then settling into a teaching life in Illinois. These first thirty years of his life account for about twenty pages in Stegner's large book. It is only when the axis of Powell's life extends east and west that Stegner takes full advantage of the storytelling promise of this life. The western end of that axis rests in the southern Rockies and the Colorado Plateau; the eastern end rests squarely in Washington, D.C. For the remainder of the biography, it is the dynamic interplay of these regions that drives the narrative.

This interplay can be seen in any of a dozen themes that Stegner develops along the east-west axis. To name just a few: the role of eastern and foreign capital in the development of the West; the role of artists and illustrators in creating the West in the public imagination; the image versus the reality of the native tribes that Powell so concerned himself with. One theme of special relevance—the post–Civil War transformation of Washington, D.C., into the nation's scientific hub—may make the point. A third of the way into the book, at the beginning of a chapter aptly titled "Center and Frontier," Stegner describes how centralization bred by the crisis of the Civil War did not cease with the peace:

> Powell himself, from 1870 on, was a forceful part of that Washington which had formed during the war and which compacted itself in the dozen years afterward. . . . Less than twenty years after the war, Washington was one of the great scientific centers of the world. It was so for a multitude of causes, but partly because America had the virgin West for Science to open, and in Washington forged the keys to open it with.[22]

The fate of government science, especially at the hands of provincial politicians from the West, then becomes the principal organizing theme for the remainder of Powell's life and Stegner's book.

It is not simply the differences between regional perspectives, but the interactions among them, that Stegner exploits in conveying the larger meaning of Powell's experience. Stegner's readers are familiar with the repeated emphasis he places on aridity as the West's defining feature—a theme we can trace back from Stegner through De Voto, Walter Prescott Webb, and Powell. But we sometimes overlook the point that it is the contrast between the dry West and the incrementally moister East, between conditions on both sides of the 100th meridian, between the traditions and social institutions that evolved to meet these conditions, that provides drama, conflict, insight, and vision. It is not the provincial mind that grasps the inadequacy of "wet weather institutions and practices" in "dry-weather country." Or vice versa. This requires continental experience and a continental vision.

Stepping back from this mountain of a book, we can see that *Beyond the Hundredth Meridian* both defined and illustrated Stegner's convictions regarding regionalism. It took someone intimately familiar with the plateau province, but not wholly of it, to do the job. We see, too, the manner in which subjects often choose their biographers—at least as much as biographers choose their subjects. Others could have written a life of John Wesley Powell; in his generation, only Wallace Stegner could have written such a biography of Powell.

The Library of Congress information on *The Uneasy Chair* captures the intermingling of experiences, influences, and styles that resulted from Stegner's decision to write a biography of Bernard De Voto. The first category is plain enough: "De Voto, Bernard Augustine, 1897–1955." Then "Authors, American—20th century—Biography." Then "Historians—United States—Biography." The lives of Stegner and De Voto had first intersected, appropriately enough, in midcontinent, at the 1937 meeting of the Modern Language Association in Chicago. (Stegner had traveled down from Madison, De Voto west from Cambridge.) They met again the following summer at the Bread Loaf Writers' Conference, and through the war years both were based in Cambridge. As De Voto took up the cudgels in the western land grab battles of the late 1940s, Stegner was writing himself more deeply into the canyonlands with Powell. If Stegner was the one person who could do full justice to Powell, De Voto was the one person who could provide a fully appreciative introduction to *Beyond the Hundredth Meridian*, which he did.[23] And in the years to come, Stegner would assume a role—conservation's literary standard-bearer—that De Voto had practically invented.

The crisscrossing of backgrounds, geographies, interests, roles, and attitudes in the De Voto–Stegner relationship would ultimately make, in 1974, for a more complex, personal, and nuanced biography than *Hundredth Meridian*. The richly stratified life presented in *The Uneasy Chair* again reflected as much on the biographer as the subject. Here Stegner was working not just from documents but from life. De Voto, like Stegner, had been a native westerner, a Utahan, a willing expatriate in the East. As De Voto's was a literary, not scientific, life, Stegner could appreciate and communicate it even more fully than he did Powell's. And Stegner wrote, not from a distance on the Inside Washington politics of science, public policy, and land law, but from within a rapidly coalescing environmental movement.

The result is an example of Edel's "third type" of biography, which combines elements of the chronicled life with more stylized portraiture, carefully sketched, placed within a sharply defined narrative frame. In this type of biography, documents are "melted down and refined so that a figure may emerge, a figure in immediate action and against changing backgrounds. Such a work tends to borrow from the methods of the novelist without, however, becoming fiction." It is "not concerned with strict chronology; it may shuttle backward and forward in a given life and seek to disengage scenes or utilize trivial incidents . . . to illuminate character."[24] In this case Stegner was as much participant as observer in the life he wrote. And to keep the austere Gods of Biography at bay, Stegner regularly injects himself into the narrative as the "skeptical" or "assiduous" biographer (perhaps a second cousin to Joe Allston and Lyman Ward).

Bernard De Voto's life-curve looped back and forth from his native Ogden, the Intermountain West, and later Montana, to Cambridge, with Harvard the dominant eastern pole, Northwestern a transition point in the middle, and important side loops encompassing New England and New York. Although the west end of the axis centered on Ogden, it shifted and expanded to include, ultimately, much of the West. The other end of the axis was fixed not, as with Powell, on the political establishment of Washington, D.C., but on the literary and intellectual establishment, the "more privileged earth," of Boston and the Northeast. Along this axis Stegner chases the central paradox of De Voto's life: because of his western upbringing, De Voto both craved and resisted the established notions of literary life as defined in the East; but he achieved the fullness of his writing only when he returned to the West as the subject of his great histories and his relentless conservation advocacy. "From early in his career," Stegner wrote, "[De Voto] had mythologized his pilgrimage eastward as a quest or trial, a journey designed to let him prove himself in the intellectual East."[25] The biography thus becomes a richly ironic exercise in demythologizing a man who himself held mythmaking in contempt and, moreover, in turning back on itself the dominant myth of the westering Euro-American.

The sharp East-West comparisons in *The Uneasy Chair* may well have resulted from the heightened awareness Stegner gained through his fictional depiction of similar circumstances in *Angle of Repose* (published three years earlier). Stegner described *Angle of Repose* as "not only a comparison of the frontier and the New West; it's a comparison of East and West. It attempts to be something relatively comprehensive about certain kinds of American experiences. . . . [Molly Foote] was quite aware of the differences between East and West, much more aware than I would be, probably, though I had some notion, having lived in both. She felt it more because she felt the West for a long time as a place of exile."[26] In a parallel manner, Stegner opens *The Uneasy Chair* with a portrait of the exile-at-home: De Voto in his early adulthood, amid personal crisis, an outcast in his own hometown of Ogden. "At best I am a spore in Utah," De Voto wrote to a friend at the time, "not adapted to the environment. . . . These people are not my people."[27]

But then De Voto proceeded to exile himself from the West. In looking at his life as a whole, it is remarkable how little time he spent in the West after leaving the region. Stegner notes that, as of 1946, De Voto's extensive knowledge of the West was "more from books than from personal experience" (p. 287). De Voto had already written *The Year of Decision: 1846*, the first installment of his great western trilogy, from a distance, and was well into the second volume, *Across the Wide Missouri*. Stegner comments tartly, "This acknowledged authority on the West was just completing a book on the fur trade without ever having set eyes on much of the country over which the fur trade had operated" (p. 288). There is something epic in the manner in which De Voto's research on the Lewis and Clark expedition led him back into the western provinces himself. In the summer of 1946—the same summer that Stegner toured the West with his family—De Voto returned West on a four-month reconnaissance and found his home.

In *The Uneasy Chair*, as De Voto's life-curve crosses back upon itself during this crucial period, Stegner as biographer plays it exactly right. Describing De Voto's return to Ogden, he writes: ". . . coming back to him on this euphoric tour of all the country he had dreamed of and made himself an expert on, his hometown looked a good deal better than it ever had before; and though he had fallen in love in Montana, he had all but forgiven Utah by the time he came across the salt desert into a Utah sunrise" (p. 293). Thus De Voto came back "to his starting point, in nostalgia instead of in derision" (p. 347). De Voto returned to Cambridge "greatly enlarged." He now had a firm sense of the outlines and relationships of western geography, a sharper understanding of how the native landscape had worked on American consciousness, and an increased confidence in the geopolitical implications of the expedition whose story he had set out to tell. The trip also initiated De Voto's full adoption of the conservation cause. Essentially it opened the spigot for his outpouring of conservation polemics. The irony, Stegner noted, was that "his crusade should have been made in behalf of the West, which he had done his best to scorn; not until he had cured himself of being literary [that is, Eastern] could he give himself back to it"(p. 322).

What makes Stegner's achievement in *The Uneasy Chair* so impressive is the seamless way in which he interweaves De Voto's exterior, geographic life and his tumultuous inner life, and reveals the connections between these worlds. Stegner was too canny to paint the inner and outer landscapes as simple reflections of one another; they aren't. What he does is demonstrate, simply, that these worlds interact in space and time and that, depending on a person's temperment and circumstance, this relationship may be at the heart of the life story. We remember De Voto's literary generation for its expatriates and its self-conscious alienation. De Voto's own foreign adventures, Stegner reports, were limited to bootlegging escapades along the Quebec border, brief auto excursions into Ontario and Alberta, and "a hypothetical trip to Mexico as a boy." Yet De Voto's view "was wide, not provincial. The perspective that others got by looking back from Europe . . . he got by looking West from Cambridge, or back from the present into the past, or forward from the past into the future" (p. 364).

STEGNER, OF COURSE, WROTE one other biography: his own. He often cautioned readers not to conflate his fiction and his life or his narrators and their author. But we also find, as in his essay "The Law of Nature and the Dream of Man," acknowledgment that "without consciously intending to, I have written my life. . . . Sure, it's autobiography. Sure, it's fiction. Either way, if you have done it right, it's true."[28] As early as *The Big Rock Candy Mountain* and as late as *Crossing to Safety*, Stegner drew upon and contrasted the regional characteristics of various Wests, California, Mexico, Iowa, Wisconsin, New England, Denmark, France, Italy, Egypt, the Philippines. To read the body of his fiction is to gain a mental map of Stegner's own travels and, as well, a sense of his striving toward unity and continuity along the way.

In a recent essay, "Finding a Voice of His Own: The Story of Wallace Stegner's Fiction," Jackson Benson has succinctly described the drawn-out and often frustrating process, unfolding over a period of more than thirty years, through which Stegner finally achieved "his true originality." In Benson's analysis, the crux of the challenge Stegner faced was "to transcend the autobiographical."[29] Thus another paradox: to find one's own voice, one must learn to speak in voices not one's own. Stegner's earliest fiction lacked the sense of personal investment that would distinguish his later works. When he turned to (thinly veiled) semi-autobiography in *The Big Rock*

Candy Mountain, Stegner began to find the distinctive, expressive tone of voice that revealed more than just raw writing talent. As Benson observes: "An author's deep participation can bring forth superior technique, and that is the great advantage of semi-autobiographical fiction."[30]

But having mined that personal field, Stegner now encountered further frustration. When his subsequent novels *Second Growth* and especially *Joe Hill* failed to connect with an audience, he left the genre behind him. During a ten-year hiatus from novels, he devoted his energies largely to short stories and to work on *Beyond the Hundredth Meridian*. Benson attributes Stegner's eventual return to novels to his working, through the short stories, toward a narrative voice that conveyed "a sense of truth and conviction coming not . . . out of the telling of his own story, but rather out of the force of personality and belief—and, one might add, his willingness to give up authoritarian control and let his characters breathe."[31] This new flexibility and vulnerability finds its first full expression in *All the Little Live Things* (1967) and continues to mature through the later novels.

What has this to do with biography and geography? Perhaps the process of composing two substantial biographies—one during his hiatus from novel-writing, the other as he was honing his new and more complex novelist's voice—helped to reorient Stegner's work generally. The biographer participates in a highly peculiar human experience: the vicarious living of another life in another time. A biographer cannot help but be enlarged, mostly unconsciously, by the experience. And if the job has been done right, the reader shares, too, in that enlargement. To a degree at least, the biographies forced Stegner to adopt new voices and perspectives, to recombine interior and exterior worlds in new ways. This may have contributed to the achievement of voice that Benson identifies. "It is a step," notes Benson, "beyond autobiographical achievement—to bring 'other' voices fully to life."[32]

Stegner insisted his interest in history was "personal, not scholarly. . . . I wouldn't have written Powell if I hadn't known the Southern Utah plateaus, and I wouldn't have written Benny De Voto's biography unless I had known him. All the history and biography I've done has been an offshoot of personal experiences and personal acquaintances."[33] But against this statement we can weigh the observation of André Maurois in *Aspects of Biography*: "Biography is a means of expression when the author has chosen his subject in order to respond to a secret need in his own nature."[34] Leon Edel notes that when the "secret need" involves the biographer's subject itself, the emotional entanglements can place the biographer in jeopardy. In Stegner's case, the need was not so secret. And it seems to have had less to do with either Powell or De Voto themselves than with the issues they confronted, the ambiguities they embodied, the work they inspired, and the insights their lives offered. Stegner did not live through his biographical subjects. He enriched his own life and literature through them, and in so doing he enriched our national life and literature.

PART III

STEGNER AS CONSERVATIONIST

STEGNER ACHIEVED RECOGNITION not only as a writer but as a conservationist. In a general sense, the past thirty years have brought to conservation a greater appreciation of its ethical and spiritual foundations, its historical and cultural context, its economic implications, and its centrality to the future not only of American society, of course, but the world. The arts and sciences underlying conservation have responded by reinventing their own approaches in a broader, more integrated fashion. The evolution of Stegner's conservation philosophy, from the 1950s on, both exemplified and stimulated these changes. Stegner became his generation's leading literary exponent of an American land ethic—an ethic reflecting a deeper sense of place and leading, in turn, to a more enduring relationship between American culture and the American land. The essays in this section trace Stegner's conservation record and its continuing consequences.

T. H. Watkins' essay "Reluctant Tiger: Wallace Stegner Takes Up the Conservation Mantle" appeared under the title "Typewritten on Both Sides" in *Audubon* magazine in 1987. Watkins was the first to summarize Stegner's contributions as "one of the central figures in the modern conservation movement." His essay is included here both for its content and as a historic document in its own right. Watkins himself writes as an active conservationist who during his years as editor of The Wilderness Society's magazine *Living Wilderness* (later *Wilderness*) regularly published Stegner.

Geographer Thomas R. Vale begins with the observation that a paradox ("or at least a puzzle") often underlies conservation challenges, especially in the American West: those who most readily express strong affection for landscapes are often the ones "most likely to criticize the character of the linkage between people and nature." Vale sees two contrasting views of the West in Stegner's writing and discovers traditional and modern elements in each. Vale argues that in Stegner's case, the paradox is resolved in "the wide arms of his embrace"—in a sense of place comprehensive enough to bracket "contrasting poles: nature and nurture, wilderness and civilization, nature preserved and nature utilized, past and present, stories of what was and what might be, the high culture of poets and the wildness of mountain water."

Conservation biologist Richard L. Knight offers a "Field Report from the New American West." For the things Stegner wrote and cared about—open space, wild places and the biodiversity they harbor, a sustainable native economy, rich local cultures, rural traditions of cooperation, and, perhaps beneath all this, a sense of modesty in what Aldo Leopold called "our land relations"—the prospects in many parts of the West are sobering. As we inherit the legacy of past environmental abuse, as the quality of life in many of our cities is allowed to deteriorate, and as wealth increases the demand for environmental amenities, the West "is filling up, once more the destination point of dreamers, boosters, and raiders." The latest waves of growth inundating the West will not ease the task of inhabitation, "the slow and painful adaptation of societies and cultures to the western environment of rugged mountains and aridity." They do, however, ensure that Stegner's testimony will remain indispensable.

Confronted by such realities, conservationists are liable to overlook their own successes in forging an American land ethic. At such times, stories are called for. In the final essay of this section, Dorothy Bradley tells a story of continuity and change, of political courage and the power of words, of the excitement that comes in forging new traditions. Above all it is a story about the acceptance of responsibility, in the tradition of Stegner. Bradley believes "there are many involved in public life in the West who, like me, found Wallace Stegner to be the translator of the land ethic for their piece of the country." It remains this constant work of citizens to translate this land ethic into reality.

Many conservationists feel an unusual affinity for Stegner not only because of the knowledge he conveyed but because he expressed the human dilemma that conservation entails. Our instincts as realists compel us to take the hard look, to see without blinking the reality of our situation, to confront without sentimentality the state of our natural world and our culture. But our commitments as citizens, scientists, parents, teachers, writers, landowners, land managers, and public officials demand that we not withdraw in cynicism or resignation. There is no easy way out. Stegner exemplified the only safe deliverance: unending reassertion of the need to recognize connections, to know our history (both human and natural), and to strengthen our sense of community (both human and natural).

CHAPTER TEN

RELUCTANT TIGER: WALLACE STEGNER TAKES UP THE CONSERVATION MANTLE

T. H. WATKINS

MORE THAN TEN YEARS have passed since I sat in a hotel room with Wallace Stegner and discussed the outlines of his conservation career for an article that first appeared in *Audubon* magazine in September 1987. (A shorter version had been published in the *South Dakota Review* a couple of years earlier.)[1] That article, as originally published, follows. In reading it again for purposes of republication, I was struck—as I had been back in 1985—by the depth and emotional character of Stegner's commitment to the land. That commitment was a visceral thing. His intellectual grounding and relentlessly logical mind were indisputably important, of course, in giving his arguments weight and clarity. But I think it was his sense of connectedness—call it love—that gave his words the power to influence the minds, hearts, and actions of those who read him.

> At the next service station where he stopped he felt it even stronger, the feeling of belonging, of being in a well-worn and familiar groove. . . . Anything beyond the Missouri was close to home, at least. He was a westerner, whatever that was. The moment he crossed the Big Sioux and got into the brown country where the raw earth showed, the minute the grass got sparser and the air dryer and the ser-

> vice stations less grandiose and the towns rattier, the moment he saw his first lonesome shack on the baking flats with a tipsy windmill creaking away at the reluctant underground water, he knew approximately where he belonged.[2]

On a hilltop above Los Altos on the rolling flanks of the Coast Range, the earth hints broadly at the fecund possibilities of a California spring. The field beyond Wallace Stegner's house has exploded in wild mustard already, and lemons and grapefruit and navel oranges ornament trees scattered among the junipers, the oaks, the eucalyptus, and the firs. Birds twitter territorial imprecations amid the tangled, greening-up chaparral of the gulches. At the fence next to the garage, a tall rosebush has burst into fat pink blossoms, too robust and too early in the season, an aberration that seems to please the novelist and his wife as they shepherd me on a walk around the property.

They are, as the saying goes, a handsome couple. Stegner is tall, still well muscled, his fine-complexioned face, the gift of Scandinavian ancestry, lined with a well-earned cynicism that is softened by the snapping blue sparkle of humor in his eyes. His wife, Mary, is small, brown, bright, expressively interested in everything around her, as if life were a gift box to be opened every day. They are both in their seventies and have fallen prey to their share and more of time's implacable erosions, but they move briskly through this spring afternoon, stopping to discuss this and that—this well-known tree, that familiar clump of flowers. I am reminded of the touching, ironic one-liner Joe Allston uses to describe himself and his wife, Ruth, each at the borderline of seventy in Stegner's novel *The Spectator Bird*: "Two young people with quite a lot the matter with us, we stood for a moment, breathing it in."[3]

Stegner wears his learning and the honors that have followed as lightly as his years, yet both are quite as weighty as those carried by anyone who has been working in the vineyard of the word over the past half-century. More weighty than most, as a matter of fact. He received his Ph.D. in literature fifty-two years ago. He has taught at the universities of Utah, Wisconsin, Harvard, and Stanford. He founded the Stanford Creative Writing Program in 1946 and directed it until his retirement as Jackson E. Reynolds Professor of Humanities in 1971. Many of the more than one hundred writers who passed through that program went on to produce enduring and even important work: Eugene Burdick, Wendell Berry, Robin

White, Robert Stone, Ken Kesey, N. Scott Momaday, Edward Abbey, Peter S. Beagle, James D. Houston, Don Moser, Harold Gilliam, Ernest Gaines, and Tillie Olson, among others. Larry McMurtry, a Stanford fellow under Stegner, won the 1986 Pulitzer Prize for fiction with *Lonesome Dove*; so did Momaday, in 1969, with *House Made of Dawn*.

In the 1940s scores of other writers were touched by his thinking at the Bread Loaf Writers' Conference, an annual event where he taught periodically beside such luminaries as Robert Frost, Archibald MacLeish, Bernard De Voto, Catherine Drinker Bowen, Louis Untermeyer, and Robert Hillyer. His own twenty-six books include twelve novels; two short story collections; two biographies; two collections of essays; three histories; the edited publication of De Voto's *Letters*; the editing of a conservation polemic (more of this later); an edited version of John Wesley Powell's *Report on the Lands of the Arid Region of the United States*, the classic 1878 study of western geographical imperatives and their human consequences; *One Nation*, an investigatory report on life in America written as World War II was drawing to a close; and, thirty-six years later, *American Places*, a collection of essays on the natural and human landscapes of the country written with his son, Page. His thirteenth novel, *Crossing to Safety*, will be published this month.

His short stories have been chosen for seven annual volumes of *The Best American Short Stories* and four volumes of the *O. Henry Awards* anthologies. He has written both fiction and nonfiction for nearly every major magazine in the United States and many of the minor ones, some of them long dead. He has been editor-in-chief of one magazine, *The American West*, and has contributed in one editorial capacity or another to several more, including *The Saturday Review*. A bibliography of his published stories, articles, critical essays, book reviews, and unclassifiable bits and pieces would run into literally thousands of items ("a lot of them best left forgotten," he says). He has won both the Pulitzer Prize (*Angle of Repose*, 1972) and the National Book Award (*The Spectator Bird*, 1977) for fiction. He has received three Guggenheim fellowships, has been a Fulbright Lecturer in Europe, Greece, and the Middle East, has taught a season at the University of Toronto, has a collection of honorary degrees, is a card-carrying member of the National Institute and Academy of Arts and Letters, the American Academy of Arts and Sciences, and Phi Beta Kappa.

The man is a walking tower of American letters, a compendium of the national literature, himself a major contributor to what he once called "the great community of recorded human experience." At the moment, however, dressed in a pastel-blue summer shirt and light-gray summer slacks, he is more the California version of the country squire, exquisitely concerned about the health and well-being of his trees, his plants, his flowers, and his fruits. He has no livestock, save for the occasional resident pussycat and, until civilization finally drove them off, a pair of coyotes that made a pretty fair living off the mice and ground squirrels and gophers around the place. He knows and tends it all intimately, affectionately, the way a good doctor knows and tends his oldest patient. He and Mary planted most of the trees themselves almost forty years ago, and all of the shrubbery and flowers—the tulips, roses, azaleas, daffodils, and others.

"Pretty good for a kid from the Cypress Hills," I say after our walk, referring to that part of Saskatchewan, Canada, where he spent most of his childhood. He chuckles and does not deny it. He has mourned his early life as rootless and wandering; born in Iowa, raised in Saskatchewan, Washington, Montana, and Utah, he lived on the thin edge always, permanence and security little more than will-o'-the-wisps. He has envied those with their feet and their traditions firmly planted in one place. So he found this hilltop in 1948 and built this house upon it and surrounded it with growing things, the root of each tree and shrub a spike nailing him down. If you can't be born to a place where you can stay, then make one.

And hold on to it. As we stand looking out over the hills from the deck at the rear of the house, a flash of irritation crosses his face. Every hill in sight has a housing compound on it. "Time was," Stegner says, "when at night you couldn't see more than two lights from here, sometimes not even that. Now there's a light everywhere you look." *Erectus americanus*, a species not exclusive to California but given elbowroom for its dreams here as in no other state, probably. For a long time Stegner has not approved of the breed and what it can do to the land. Sabrina Castro, the heroine of his 1961 novel *A Shooting Star*, was not speaking independently of her maker when she ruminated about her brother, the real estate developer:

> His kind never anticipated consequences. His was the kind that left eroded gulches and cutover timberlands and man-made deserts and jerry-built tracts that would turn into slums in less than a generation.... They denuded and uglified the earth in the name of progress, and when they lay in their deathbeds—or dropped from the massive coronary that the pace of their lives prepared for them—they

> were buried full of honors and rolling in wealth, and it never occurred to the people who honored them, anymore than it had occurred to themselves, that they nearly always left the earth poorer and drearier for their having lived on it.[4]

The anger is muted now but not gone. Probably it never will be gone. For the emotion from which it springs is too firmly planted, spiked into the land like the trees that mark his own patch of ground in California, mark it and pin him to it. "West is a country in the mind, and so eternal," Archibald MacLeish wrote. It certainly has been a country firmly in the mind and work of Wallace Stegner. Just as surely as the western experience has shaped us as a nation, it has informed the spirit and intellect of Stegner's writing. It is neither a casual nor a superficial influence; it has precious little to do with cowboys and even less with Indians—at least as representative myths. It runs deep, down to the marrow where imagination lies, and is all of a piece with the man.

No other major modern writer of fiction has known so much of the West from personal experience, none has so steeped himself in its history, and none has done so much with what he found in both. "As a regional writer," Wendell Berry wrote in the winter 1985 issue of the *South Dakota Review*, "he seems to me exemplary."

> He has worked strenuously to know his region. He has been not just a student of its history, but one of its historians. There is an instructive humility in his studentship as a historian of the West. . . . He has the care and scrupulousness of one who understands remembering as a duty, and who therefore understands historical insight and honesty as duties. He has endeavored to understand the differences of his region from other regions and also from its own pipedreams and fantasies of itself. He has never condescended to his region—an impossibility, since he has so profoundly understood himself as a part of it. He has not dealt in the quaint, the fantastical, or the picturesque. And, above all, he has written well.[5]

Filtered through intelligence and the gift of vision and language, the regional then becomes the universal. And so, some would say (myself among them), eternal.

Out of that same personal experience, intellectual inquiry, and creative expression there has emerged a powerful, proprietary love of the land itself, a love that gives this man's life and work an even larger dimension. It has, in fact, made him one of the central figures in the modern conservation movement. I say this from the safety of my study in Washington, D.C. If I accused him of it to his face, he would box his ankles, wave an arm in dismissal of the whole idea, and maybe aim a kick at my shin. He would point out that we have known each other for twenty years and that I am therefore biased, prejudiced, necessarily, unavoidably myopic; he would mutter about historical objectivity and critical decorum. Nevertheless, what I say is true. With writing and teaching there has been a third career that has divided Stegner's time, talent, and devotion. (There is no better word.) For forty years he has borne witness for the land that has enriched his life and art, and the measured cadence of his splendid prose has played a significant role in the shaping of the sensibility we now call environmentalism.

> THE DRAMA OF THIS LANDSCAPE is in the sky, pouring with light and always moving. The earth is passive. And yet the beauty I am struck by, both as present fact and as revived memory, is a fusion: this sky would not be so spectacular without this earth to change and glow and darken under it. And whatever the sky might do, however the earth is shaken or darkened, the Euclidean perfection abides. The very scale, the hugeness of simple forms, emphasizes stability. It is not hills and mountains which we should call eternal. Nature abhors an elevation as much as it abhors a vacuum; a hill is no sooner elevated than the forces of erosion begin tearing it down. These prairies are quiescent, close to static; looked at for any length of time, they begin to impose their awful perfection on the observer's mind. Eternity is a peneplain.[6]

We didn't call the movement environmentalism in the very beginning, of course. When we thought about it at all—and most Americans did not—we called it conservation. And forty years ago it was by no stretch of the imagination the relatively coherent thing it has since become. It was, more accurately, a sentiment, one variously held and variously expressed by a handful of organizations that ranged from a comparatively militant National Parks Association to a firmly benign Sierra Club. With the exception of occasional twitches—Bob Marshall forming the Wilderness Society, Rosalie Edge scolding the National Audubon Society into advocacy, that sort of thing—the first lurch of the movement did not come until the 1940s. For in the years preceding that decade the common assumption remained that progress as we defined it in this country was an unrelieved good and that land put to human purposes was fulfilling its highest meaning.

Stegner himself had spent much of his life accepting this notion. "I guess I was pretty innocent," he says, "because I hadn't ever put my mind to problems, and all during the war, of course, we were shut off by gas rationing from getting out into the country to see what was happening. Before the war, from 1937 on, I was in the Midwest and East—at Wisconsin and Harvard—where the problems of public land, environmental land use at least, didn't come up and where land actually heals faster than in the West. And before that, I was at the University of Utah as a student when Hoover Dam was being planned and we all thought it was a great geology excursion to go down and look at Black Canyon, where the dam would be put in. I guess I just accepted the habits of mind of most of the people of Utah at the time that this was progress—lots of jobs. The Utah Construction Company hired all kinds of people, and I used to play ping-pong with the son of the head of the company. All of these influences—none of them intellectual, none of them historical, none of them environmental—were working on me."

In 1946, the war over and gas rationing done, the Stegners, with son Page and another couple, went out into the West again to see what they could see. Among the sights was Hoover (then Boulder) Dam, completed now for ten years, a celebration in concrete that had driven poet May Sarton, among others, to ululations of praise:

> Not built on terror like the empty pyramid,
> Not built to conquer but illuminate a world:
> It is the human answer to a human need,
> Power in absolute control, freed as a gift,
> A pure creative act, God when the world was born,
> It proves that we have built for life and built for love
> And when we are all dead, this dam will stand and give.[7]

There was still innocence enough left for Stegner to be infected, too. "Nobody can visit Boulder Dam itself without getting that World's Fair feeling," he wrote in an article for the *Saturday Review* that year: "It is certainly one of the world's wonders, that sweeping cliff of concrete, those impetuous elevators, the labyrinth of tunnels, the huge power stations. Everything about the dam is marked by the immense smooth efficient beauty that seems peculiarly American."[8]

Yet even as he batted these words out on his Remington, something else was at work on him. For several years now, following an impulse that had come over him while teaching at the University of Utah from 1934 to 1937, he had been gathering materials for a biography of John Wesley Powell, geologist and explorer, commander of the first expedition to descend the Colorado River through the Grand Canyon in 1869, founder of the U.S. Geological Survey and the Bureau of American Ethnology in 1879, and author of what is still considered one of the most percipient documents ever to tumble from the presses of the Government Printing Office: the 1878 *Report on the Lands of the Arid Region of the United States.* In cool, analytical style, Powell had taken on all the boomers and boosters who were trying to make the arid and semiarid West what it could never be, the new garden of the world, stuffed with immeasurable resources and capable of supporting a limitless population of happy and productive citizens. Not so, said Powell: all the West's resources were finite, and the most precious resource of all—water—was finite to such a degree that only through careful planning and land stewardship could any kind of civilization flourish between the 100th meridian on the east and the upthrust of the Sierra Nevada on the west.

This was not then (nor is it now, in many quarters) a popular view, merely an intelligent one, and it had begun to color Stegner's thinking. Before Powell, he says, "there was a lot of country that I knew, but I didn't have any wits about it." He did by the end of 1946, when Wyoming Congressman Frank Barrett, a sometime livestock rancher and lawyer out of Lusk, decided that it would be a splendid idea to have the feds turn over all government grazing land in the unreserved public domain and national forests to the individual states for eventual disposal to the cattle and sheep industries. Barrett had been in an almost continual snit since 1943, when Franklin D. Roosevelt, at the urging of his curmudgeonly secretary of the interior, Harold Ickes, had created the Jackson Hole National Monument beneath the towering magnificence of the Teton Range. Quite rightly, Barrett and the industry he represented saw this "federal land grab" as an attempt to protect this beautiful stretch of country from overgrazing. And when repeated efforts to get the monument rescinded failed him, he determined to initiate a territorial grab of his own by introducing legislation that would have divested the government of all its grazing lands.

Frank Barrett's grandiose scheme galvanized the conservationists, including nascent environmentalist Wallace Stegner, as nothing had for decades. As Stegner remembers it, his involvement came through the direct intercession of another born westerner who had just returned from a homecoming trip that year of 1946—the Cambridge-based Bernard De Voto, Stegner's old acquaintance from the Bread Loaf conferences, who numbered among his wide scattering of disciplines proprietorship of "The Easy Chair" column in *Harper's* magazine. During his own trip, De Voto had caught wind of the stockmen's gambit and upon his return had penned an "Easy Chair" piece on the subject that caused Stegner to write him in approbation: "I expanded upon my 'hurrays' in some fashion to blow off something I had noticed, and he wrote back at once, saying, 'If you feel like that, goddammit, get into print with it—we need everybody in print that we can get.' So I did, dutifully; my guru had spoken." The resulting article went to *The Reporter*, and from that point forward Stegner became a dedicated foot soldier in De Voto's small army of freelance conservation polemicists—a force that included at one time or another such writers as Struthers Burt, Stewart Holbrook, A. B. Guthrie Jr., William Vogt, J. Frank Dobie, and even national columnists Elmer Davis, Marquis Childs, and Joe Alsop.

The public pressure this media hit team was able to generate went a long way toward assuring the demise of Congressman Barrett's spectacular ambitions. But it was only the opening skirmish in a new campaign of the great American conflict of values, one whose history could be tracked back to the middle of the nineteenth century and the battle cry of the gentle Henry David Thoreau: "In wilderness is the preservation of the world." Much of the generation of Americans who emerged unscathed from World War II could not understand that kind of thinking, including the staffs of the Bureau of Reclamation and the Army Corps of Engineers, who had a tradition of looking upon western rivers and finding them good—for dams. Big dams.

Two of the biggest soon appeared as plans for Split Mountain and Echo Park, on the Green River as it flows through Dinosaur National Monument—a piece of plateau country bisected by the Utah-Colorado border. Stegner knew the country and the river well; it was on the Green, after all, that Powell and his crew had started out in 1869, and by now Stegner was well into the writing of *Beyond the Hundredth Meridian: John Wesley Powell and the Second Opening of the West.* He spent the summer of 1948 at Struthers Burt's Three Rivers Ranch in Jackson Hole, working on the book and listening attentively as Burt, Arthur Carhart of the Izaak Walton League, and De Voto poured out their vigorous conviction that either dam would flood significant portions of the monument to no good end whatever. By the end of the summer, he says, "I was pretty well back into the thick of things."

So were a lot of people. The controversy did not reach fever pitch until 1950, when Interior Secretary Oscar Chapman gave official approval to the dams as part of the Upper Colorado River Basin Storage Project, but after that it took on dimensions that justify its description as the central event in the narrative of modern conservation. As Irving Brant of the Emergency Conservation Committee noted at the time: "For the first time in the history of conservation . . . practically all organizations devoted to it have been stirred up to a joint campaign." There were seventeen groups finally (including the National Audubon Society), led by David Brower of the Sierra Club, Howard Zahniser and Olaus Murie of the Wilderness Society, Ira Gabrielson of the Wildlife Management Institute, and William Voigt Jr. of the Izaak Walton League. Arrayed on the other side were the formidable lobbies of the Bureau of Reclamation and the Army Corps of Engineers, Truman's Interior Secretary Chapman, and after him Eisenhower's man, Douglas McKay, a former automobile salesman from Oregon. The battleground was Congress. Legislation for approval of the Upper Colorado Project was pending.

In articles for *The Reporter* and *The New Republic* Stegner added to the furious blizzard of letters, articles, and testimony thrown in the face of Congress by agitated conservationists. In his own contributions, he presented the argument against the dams with the cool precision typical of his orderly mind; but he also was angry, and sometimes it showed. Stegner wrote in *The New Republic:*

> The fact that Reclamation has picked these sites within the Monument, and plugged for them, suggests that perhaps it *wants* to infringe on the sanctity of the parks. Almost every western stream of importance touches, in itself or its tributaries, some park or monument; and Reclamation has the job of taming every western stream. It could

> build dams with a much freer hand and have to compromise less with other interests if it could break down the national park immunity. Something similar might be guessed of Secretary McKay: one who ponders the evidence might well conclude not only that Secretary McKay is willing to violate park territory, but he would like nothing better.[9]

In 1954 Alfred A. Knopf agreed to publish a book in support of the cause, and the Sierra Club's David Brower, who had the face of a child but the heart of a Druid, came down to Stanford and persuaded Stegner to edit the book. The result—a labor of love for all concerned—was *This Is Dinosaur: Echo Park and Its Magic Rivers,* with chapters by Stegner, Olaus Murie, and others, including Knopf himself. It was published at breakneck speed in 1955 and sent to every member of Congress.

The accumulated pressure was too much. When the project finally was passed, it included a stipulation that no national park or monument would be violated by any of its parts. The Echo Park and Split Mountain dams were dead.

▽ ▽ ▽

> THROUGH MOST OF ITS COURSE the canyoned Green and Colorado, though impressive beyond description, awesome and colorful and bizarre, is scenically disturbing, a trouble to the mind. It works on the nerves, there is no repose in it, nothing that is soft. The water-road emphasizes what the walls begin: a restlessness and excitement and irritability. But Glen Canyon . . . is completely different. As beautiful as any of the canyons, it is almost absolutely serene, an interlude for a pastoral flute. Except for some riffles in the upper section its river is wide, smooth, deep, spinning in dignified whirlpools and moving no more than seven or eight miles an hour. Its walls are the monolithic Navajo sandstone, sometimes smooth and vertical, rounding off to domes at the rims, sometimes undercut by great arched caves, sometimes fantastically eroded by slit side canyons, alcoves, grottoes green with redbud and maidenhair and with springs of sweet water.[10]

"In a realpolitik sense, the movement came of age during Echo Park," the historian Stephen Fox has written. "The conservationists, perhaps to their own surprise, beat down powerful federal bureaus and private commercial interests."[11] If so, surprise ultimately was mixed with sorrow, for the victory (like most conservation victories) came attached to a compromise: to eliminate the Echo Park and Split Mountain sites, the conservationists had offered up Glen Canyon on the Colorado as an alternative. The Bureau of Reclamation seized upon the idea and got to work. The result was Glen Canyon Dam, dedicated in 1965, and behind it Lake Powell, which buried the entire beautiful trench from which the dam had taken its name. "We could have strangled it (the Upper Colorado Project) to death," Stegner told Richard Etulain in *Conversations with Wallace Stegner*:

> We had it down completely. They took the dams out of Dinosaur in a desperate conviction that if they didn't they were going to lose the whole thing. Maybe we should have been harder-nosed than we were. . . . Having saved Dinosaur, we accepted the ruin of Glen Canyon, which was not very smart of us. Dave [Brower] has regretted that all his life. Nobody knew Glen Canyon then except me; I'd been down it a couple of times and I told him it was better than Echo Park. He didn't believe it, and I didn't push it. But anyway, that first book got me into the thing. . . . I kept getting mad.[12]

He got better than mad. He got productive, and there was plenty to be productive about. Wilderness, for example. Having gathered to defend Dinosaur in 1950, the movement by the end of the decade had capitalized on cooperation by going on the offensive, and in 1957 Senator Hubert H. Humphrey introduced the first wilderness bill, legislation written largely by Howard Zahniser. For the next seven years this attempt to create a nationwide system of protected wild country defined the sophisticated purpose of the new conservation. The idea had plenty of supporters (and plenty of enemies) in and out of Congress. What it lacked was a manifesto, a central document, what Stegner himself would come to call a coda. On December 3, 1960, in a letter to David E. Pesonen of the Outdoor Recreation Resources Review Commission, Stegner provided it. It was, he has said, "the labor of an afternoon," but the ripples from this pebble in a pool have traveled far. Even before it could be published as part of the commission's 1962 report, *Outdoor Recreation for America*, the letter had been used by Interior Secretary Stewart Udall as the basis of a speech he gave to a wilderness conference and had been published as the centerpiece from that conference. *The Washington Post* picked it up and printed it whole; it was found pasted on the wall of an office in a Kenya game park; all or parts of it have appeared on posters in Zimbabwe, Canada, and Australia; its last four words, "the geography of hope," were used as the title for a book of Eliot Porter photographs; conservation writers (myself included) have stolen from it shamelessly and quoted it endlessly.

"I take this as evidence not of special literary worth," Stegner wrote twenty years after the letter itself, "but of an earnest, worldwide belief in the idea it expresses."[13] True, the idea has power; but so do the words he chose to present it. It is the statement of a man who has thought deeply about what he is going to say, and his manner of saying it goes a long way toward demonstrating why Stegner's work rises up from the semiarid plains of our literature like the Front Range of the Rockies. Here is just a piece from the end of it:

> It is a lovely and terrible wilderness, such a wilderness as Christ and the prophets went out into; harshly and beautifully colored, broken and worn until its bones are exposed, its great sky without a smudge of taint from Technocracy, and in hidden corners and pockets under its cliffs the sudden poetry of springs. Save a piece of country like that intact, and it does not matter in the slightest that only a few people every year will go into it. That is precisely its value. Roads would be a desecration, crowds would ruin it. But those who haven't the strength or youth to go into it and live can simply sit and look. They can look two hundred miles. . . . They can also look as deeply into themselves as anywhere I know. . . .
>
> These are some of the things wilderness can do for us. That is the reason we need to put into effect, for its preservation, some other principle than the principles of exploitation or "usefulness" or even recreation. We simply need that wild country available to us, even if we never do more than drive to its edge and look in. For it can be a means of reassuring ourselves of our sanity as creatures, a part of the geography of hope.[14]

The Wilderness Act was passed by Congress and signed into law by President Lyndon B. Johnson in September 1964—a stroke of a pen that designated 9 million acres of wilderness areas on the American public lands. (The National Wilderness Preservation System has since grown to 88.6 million acres.) By this time, Stegner was in the thick of things indeed. He had been appointed to the Sierra Club's publications committee shortly after *This Is Dinosaur* was issued. With such others as August Frugé, Martin Litton, and Paul Brooks, he gave Brower the support necessary to produce a publishing phenomenon of the 1960s—the "Exhibit Format" series of handsome photographic books whose purpose was to display the beauty and inspire the protection of wild country. (The series also included a eulogy to the vanished Glen Canyon, *The Place No One Knew.*) Because of the Dinosaur book, Stegner was made an honorary life member of the Sierra Club, and Ansel Adams, one of the Old Turks of the organization, talked him into running for election to the board. He won, and served for two years.

The Sierra Club was by no means the only organization to which he was tied in these activist years. Having committed himself to the cause, Stegner spread out in any number of directions. He even helped to invent an organization. In the fall of 1960 he became a founder and honorary president of a group of local citizens called the Committee for Green Foothills, which exerted pressure on foothill communities to limit the frenzied development then, as always, prevalent in most of California. "We were trying to save them from county carelessness," he says. "The county seat was way down in San Jose, a long way off, and nobody gave a damn about the foothills down there. The developers were doing pretty much as they pleased. We operated primarily to keep a sharp citizen's watch on planning commissions and town council meetings."

On the statewide level, Stegner joined the advisory board of Alfred Heller's California Tomorrow, a muscular little outfit devoted to goading the state's government into producing a land-use plan with teeth. It failed in that endeavor, but it was instrumental in forcing passage of Proposition 20 in 1972—an act that created the California Coastal Commission to control the development and use of the state's twelve hundred miles of seashore. On the national level, he served on the councils of the Trust for Public Land and People for Open Space. Like his conservation mentor, Bernard De Voto (who had died in 1955), he was appointed to the Advisory Board for National Parks, Historical Sites, Buildings, and Monuments, a citizens' committee whose function was to advise the National Park Service on questions of park expansion, protection, and use. (It is now defunct, done in as part of the Reagan administration's quest for fiscal responsibility.) He served from 1962 to 1965 and in 1965 was the board's chairman.

It was not his first brush with official Washington in matters of conservation, for in 1961 he had worked as a special assistant for Interior Secretary Stewart Udall—although it took considerable persuasion on Udall's part to haul him into the rabbit warrens of bureaucracy. "When I heard that President Kennedy had appointed him secretary of the interior," Stegner remembers, "I was sufficiently aware of what the situation was so that I cheered. I thought it was time we had a good secretary of the interior; so I sent Udall a copy of *Beyond the Hundredth Meridian*, with a little note saying, 'Hooray, I am glad to see you in there.' This was not very characteristic of me, but I was enthusiastic." So was Udall. After a subsequent meeting during a speaking tour Stegner undertook for Phi Beta Kappa in the spring of 1961, Udall asked him to work for him as a consultant, particularly in matters of park expansion in the plateau country. Stegner pleaded an abundance of commitments and a dearth of time, but Udall would not give up. By the end of the summer, Stegner had caved in. Juggling time and responsibilities, he managed to carve out at least a few months for Udall.

One of the first things Udall did was to send his new assistant on an expedition to the country of his adolescence and young manhood—the canyoned and rock-ribbed landscape of southern Utah. Stegner, fellow employee Joe Carrithers, and a crew of interested citizens set out by car from Moab, through Capitol Reef National Monument, down the Kaiparowits Plateau to Hole-in-the-Rock, and by horseback into the Escalante Basin, where they were among the last human beings to view the magnificent Cathedral-in-the-Desert before the waters of Lake Powell buried it forever. They then climbed over the Aquarius Plateau to Capitol Reef again, where they spent three days casing the country around the monument to see what additions would be necessary to give it national park status, one of Stegner's hopes. He and Carrithers returned to Washington with the recommendation that Capitol Reef become a national park. They had a fistful of other ideas as well. Among them was the notion that the Bureau of Land Management, which administered most of the land they had just explored, should designate major portions of it as official wilderness areas—similar to the "primitive area" classifications the Forest Service had placed on some of its own holdings even before passage of the Wilderness Act. (Capitol Reef was enlarged and did become a national park ultimately, but the BLM is still dragging its feet over wilderness recommendations for much of the country Stegner's party traversed thirty years ago.)

By then, Stegner says, "Stewart was deep into the notion of a book which would express the whole philosophy of saving land for public purposes. So he put me on that project for the last month or so of my tenure there. I spent a month in the Library of Congress, reading and putting together an outline of what such a book might contain, and one of the last things we did was to go over that outline." The result was *The Quiet Crisis*, published in 1963. With Rachel Carson's *Silent Spring*, issued the year before, the book became one of the talismans of the resurgent conservation movement. Stegner is careful to point out that although he and others—his former student Don Moser; Harold Gilliam, another former student; historian Alvin Josephy—may have drafted what he calls "pilot" chapters, "it was very much Udall's book. He put his own stamp on it."

Stegner came away from his tour of duty at Interior much impressed by Udall, whom he considers one of the best secretaries in U.S. history: "He had a real feel for saving country and, when the chips were down, the nerve to take very unpopular political stands, including being against proposed dams in the Grand Canyon, which meant he had no political future in Arizona. As a steward of the land, I would rate him very high indeed." Coming as it does from an entrenched westerner, a respected historian, and a longtime participant in the conservation movement, this is probably a statement Stewart Udall can learn to live with.

Characteristically, Stegner gives himself poor marks as a conservation advocate. "I have not been an effective or even eager activist," he once wrote me. "In all the issues that matter, there are dozens of people—Dave Brower, Ed Wayburn (of the Sierra Club), Howard Zahniser, the hard-nosed, tough, and durable types who run the Trust for Public Land, the American Farmland Trust, and the big conservation organizations—who have had an immediate, practical, effective usefulness. I never have. . . . Actually I would like, and would always have liked, nothing better than to stay home and write novels and histories, and when the compulsions of some book get too strong for me, I have a history of backing away from environmental activism. . . . I am a paper tiger, Watkins, typewritten on both sides. Get that in somewhere."

Okay. But another opinion was expressed in 1982 when the Sierra Club gave him its prestigious John Muir Award.

EVENTUALLY THEY CLIMBED through the clouds and into the sun. Looking back, Susan saw South Park filled nearly to the brim with cloud, only the saw-toothed peaks rising above it. . . . They came out onto a plateau and passed through aspens still leafless, with drifts deep

> among the trunks, then through a scattering of alpine firs that grew runty and gnarled and gave way to brown grass that showed the faintest tint of green on the southward slopes and disappeared under deep snowbanks on the northward ones. The whole high upland glittered with light. . . .
>
> The skyline, from any part of this magical plateau, was toothed like the jaw of a shark. The road bent and dipped down through a hanging valley where mosquitoes rose in swarms from the wet grass; when it lifted them again around a corner of bare stone the mosquitoes blew away instantly, and the wind was so cold it made her teeth ache. Her eyes watered with cold and light. . . .
>
> The thin air smelled of stone and snow, the sun came through it and lay warm on her hands and face without warming the air itself. Up, up, up. There was no top to this pass. . . . They were long past all trees, even runted ones. The peaks were close around them, the distance heaved with stony ridges, needles, pyramids in whose shadowed cirques the snow curved smoothly. The horses stopped, pumping for air, and as they rested she saw below a slumping snowbank the shine of beginning melt, and in the very edge between thaw and freeze a clump of cream-colored flowers.[15]

All during the years of private and public service in the cause of the land, Stegner, the "paper tiger," never stilled his voice in print. He is still typewritten on both sides, and it is the writing that will define the final dimensions of his contribution to the movement. When De Voto died, Stegner reported in *The Uneasy Chair*, his biography of the one-man artillery regiment, "editorial writers and politicians and conservationists memorialized him and regretted his passing and wondered who would do his work. They were agreed it would take three men."[16]

Stegner himself assumed much of the burden for conservation. And if his deliberate instincts produced less explosive prose than De Voto, the stately, resonant sentences that came out of him had a power all their own. In 1964, for instance, he went down into the Escalante Basin again to see what had happened since Lake Powell began filling up behind Glen Canyon Dam. He found nothing much there he liked, but just a little beyond the lake there was country that he still would have had us save:

> I have been in most of the side gulches off the Escalante—Coyote Gulch, Hurricane Wash, Davis Canyon, and the rest. All of them have bridges, windows, amphitheaters, grottoes, sudden pockets of green.

> And some of them, including the superlative Coyote Gulch down which even now it is possible to take a packtrain to the river, will never be drowned even if Lake Powell rises to its planned thirty-seven-hundred-foot level. What might have been done for Glen Canyon as a whole may still be done for the higher tributaries of the Escalante. Why not? In the name of scenery, silence, sanity, why not?[17]

Why not, indeed, was the reaction such statements were designed to elicit—and, from many, did. In the pages of *The Atlantic, Harper's, The Saturday Review, Blair & Ketchum's Country Journal, The New Republic, Esquire*, and *American Heritage*, as well as the conservation press, and in speeches and book reviews and introductions, he hammered the message home: we had allowed ourselves to become prisoners of our history and were threatening to obliterate the land.

"The potentials that bring a gleam to the eye of entrepreneurs may frighten others," he wrote in the introduction to his 1969 collection of essays, *The Sound of Mountain Water: The Changing American West*,

> for all of the West's resources, even water, even scenery, are more vulnerable than the resources of other regions, and, perhaps as a consequence of that fact, the social and economic structure of the West is tentative, uncertain, and shifting. Short-haul freight rates and distances from sizable markets are a handicap, but a more compelling fact is that the basic resources of water and soil, which can be mismanaged elsewhere without necessarily drastic consequences, cannot be mismanaged in the West without consequences that are immediate and catastrophic. . . . And the entire history of the West, when we hold at arm's length the excitement, the adventure, the romance, and the legendry, is a history of resources often mismanaged and of compelling conditions often misunderstood or disregarded. Here, as elsewhere, settlement went by trial and error, only here the trials were sometimes terrible for those who suffered them, and the errors did permanent damage to the land.[18]

However clearly he saw his native country, he was then still unwilling to dismiss its capacity to learn. "Angry as one may be at what heedless men have done and still do to a noble habitat," he concluded in that same essay, "one cannot be pessimistic about the West. This is the native home of hope. When it fully learns that cooperation, not rugged individualism, is the quality that most characterizes and preserves it, then it will have achieved itself and outlived its origins. Then it has a chance to create a society to match its scenery."[19]

Stegner allowed that statement to stand untouched in the October 1985 printing of *The Sound of Mountain Water*, but an edgy note had begun to creep into some of his prose. In an essay on Aldo Leopold's land ethic for the spring 1985 issue of *Wilderness*, he seemed a little less certain about our chances:

> There will always be those who believe that the public domain is owned and maintained by the rest of us for their exclusive exploitation and use. There will always be those who never breathed deeply of the wind off clean grassland, or enjoyed the shade of a hillside oak on a July day, or watched with pleasure as some wild thing slid out of sight, and who will never see the reasons for living with the earth instead of against it. The number of functional illiterates that our free public education produces does not make us sanguine about educating the majority of the public to respect the earth, a harder form of literacy. Leopold's land ethic is not a fact but a task. Like old age, it is nothing to be overly optimistic about. But consider the alternative.[20]

Shoeless on a hotel bed in Washington, D.C., where he had come in March 1985 to attend the semiannual meeting of the Wilderness Society's Governing Council (he had joined the council in 1984), Stegner told me why a certain reserve had clouded his outlook between 1969 and 1985. "I think those sixteen years have taken some skin off," he said with a wry chuckle. "I think it's natural for people who live in what you might call a burgeoning nation-society to be optimistic, because you're always building something. But you watch what's happened to what you're building, and I think you get a little jaundiced. What's happened, of course, is that big—enormous—and quite irresistible *money* has taken over the West. And resources—extractable resources—are what that money is after."

The powerful infusion of corporate capital, he said, has all but overwhelmed the smaller economic structures of the West—and with them the very social fabric of the towns in which they function, not to mention the style of living that has always been tied directly to the land and its continuing good health. "The curious thing," he said, "is that the resistance to that, or at least some part of it, comes not from the natives, who you'd think would know best what their own interest is, but from people who have been brought into the towns partly by the corporate invasion of money—young, educated city people who want something simpler. The conservation groups that you run into out there—the ones I go talk to—are very largely made up of such young people. It's odd that the invasion of corporate cash, corporate greed, should bring not only the danger but also, and almost at once, a kind of population of resistance."

It is in these young people and in the legislative process—our system of laws—that Stegner could still discern hope. "I think everything of importance that's happened in the environmental movement has come about through the passage of laws," he said. "You can't do anything unless you have a law to back you up. We couldn't have stopped the dams in Dinosaur if it hadn't been for the National Park Service Act, which said the parks were to be enjoyed without impairment. The minute you have even a law that a lot of people gnashed their teeth over, like the Endangered Species Act, you can save something—you can save San Bruno Mountain (in Northern California) because of two little species of butterflies that exist nowhere else." He laughed. "I'm going to sound hopeful again in a minute. But it's kind of like a stubborn child: over a period of time he may leave home on account of what you tried to teach him, but when he finds himself married and raising kids of his own, he echoes what you told him. It's possible to educate a public in that way—sometimes you don't even know that it's been done."

If it has been done, it very largely has been done because we stubborn children, or at least a lot of us, have learned from Stegner's own vision and from the words that have spilled out of his mind and heart with such clarity, such presence, such power, for the past forty years. They are the gift of a white-haired genie sitting on a California hilltop muttering ancient Norwegian curses against all those who would mortgage the future of the land, the western land, the American land; his land, his native land.

> WE ARE THE MOST various people on earth, and every segment of us has to learn the lessons both of democracy and of conservation all anew. . . . What freedom means is freedom to choose, but it takes a long time to learn how to choose, and between what options. If we choose badly, we have, not always intentionally, violated the contract. Democracy assumes, on the strength of the most radical document in human history, that all men are created equal, and that given freedom

they can become better masters for themselves than any king or despot. But until we arrive at a land ethic that fits both science and human affections, until we achieve some common reverence for the earth that has blessed us, Americans are likely to be what Aldo Leopold in an irritable moment called them: people remodeling the Alhambra with a bulldozer, and proud of their yardage. . . .

It would promise us a more serene and confident future if, at the start of our sixth century of residence in America, we began to listen to the land, and hear what it says, and know what it can and cannot do.[21]

CHAPTER ELEVEN

NATURE AND PEOPLE IN THE AMERICAN WEST: GUIDANCE FROM STEGNER'S SENSE OF PLACE

THOMAS R. VALE

> When Bruce drove west in June . . . he drove directly from rainy spring into deep summer, from prison into freedom. . . . He was a westerner, whatever that was. The moment he crossed the Big Sioux and got into the brown country where the raw earth showed, the minute the grass got sparser and air dryer and the service stations less grandiose and the towns rattier, the moment he saw his first lonesome shack on the baking flats with a tipsy windmill creaking away at the reluctant underground water, he knew approximately where he belonged. . . . At sunset he was still wheeling across the plains toward Chamberlain, the sun fiery through the dust and the wide wings of the west going red to saffron to green as he watched, and the horizon ahead of him vast and empty and beckoning like an open gate. At ten o'clock he was still driving, and at twelve. As long as the road ran west he didn't want to stop, because that was where he was going, west beyond the Dakotas toward home.[1]

The description of Bruce Mason's homeward journey across South Dakota is one of my favorite passages in Wallace Stegner's writing. It moves me emotionally not only for its evocative image of the landscape, the visual scene, but also for its strong sense of place, the region all of us call the

American West. Like Bruce, and Stegner himself, my own emotional involvement with places western derives from the contingency of birth: "I came from the arid lands, and liked where I came from."[2]

The concept of place is much studied and evaluated in the thoughtful scholarly literature. Usually the idea involves a constellation of related elements: nature and people interact to create a landscape; a person, through long association with that landscape, becomes familiar with it; that familiarity breeds affection; with that positive, emotive bond, the landscape becomes a place. The scale of locales for which "placeness" may be felt, moreover, varies widely—from the comfortable chair in the living room to a neighborhood of several city blocks to a region as vast as the Deep South or New England.

Applied to the American West, however, the concept of place presents a dilemma, a paradox, or at least a puzzle, because, in their common western manifestation, the interacting elements that together define "place" seem contradictory. On the one hand, those persons who most readily identify with the West hold the region and its landscapes with the affection that is central to "placeness." On the other hand, these same people are the very ones most likely to criticize the character of the linkage between people and nature in the West and to regret what humans, through their long association with western landscapes, have done to the region.

Stegner himself represents the problematic application of the idea of place, as commonly articulated, to the West. In his own words, he felt that he had "an exaggerated sense of place."[3] And it was the dry lands of the West where his sense was focused: "I was used to a dry clarity and sharpness in the air. I was used to horizons that either lifted into jagged ranges or rimmed the geometrical circle of the flat world. I was used to seeing a long way . . . [the] smell of sagebrush, and the sight of bare ground."[4] And yet his early life was one of constant motion—motion within the West, but motion nonetheless—and such movement, Stegner suggests, does not permit the development of the familiarity with locale that is the prerequisite for feeling placed: "A place is not a place until people have been born in it, have grown up in it, lived in it, known it, died in it—have both experienced it and shaped it . . . over more than one generation" (p. 201). Stegner's own life thus exemplifies the paradox of conflicting interacting elements: affection, yes; but rootedness in a specific situation for more than a generation, no.

What is the nature of this dilemma? How might we resolve it? The question might be stated more simply: what does it mean to love the West as a place? Or: what is the basis of Stegner's sense of place in the West?

People and Nature in the West

Clearly, Stegner loved the West. His writings, particularly his fiction, may occasionally have been set elsewhere—in the Midwest, New England, even in Europe—but the bulk of his attention was directed toward that part of the world he knew best and felt most strongly about: the American West. The themes woven through his novels, stories, histories, and essays always extended beyond the regional focus—Stegner had a continental, often a universal, vision of the human condition—although he chose to write of individual lives, interactions among people, and ties between humans and nature within the context of the Great Plains, the Interior West, and lowland California. It is an accurate label, not a pejorative one, to say that Stegner was a regional author.

Stegner's vision of the West as a region focused on a broad overlapping of the human and natural worlds but not a complete coincidence of the two realms; nature was, for Stegner, distinct from humanity (Figure 1). But it is where these realms overlap that we can locate the characteristic themes in his writing, themes that allow us to define the character of his sense of place (Figure 2).

Two sets of themes dominate Stegner's writing. The more common set looks backward in time, to the past, in which the present is explained by the interactions among the specific components of the human and natural worlds through the period of the initial westering of American society

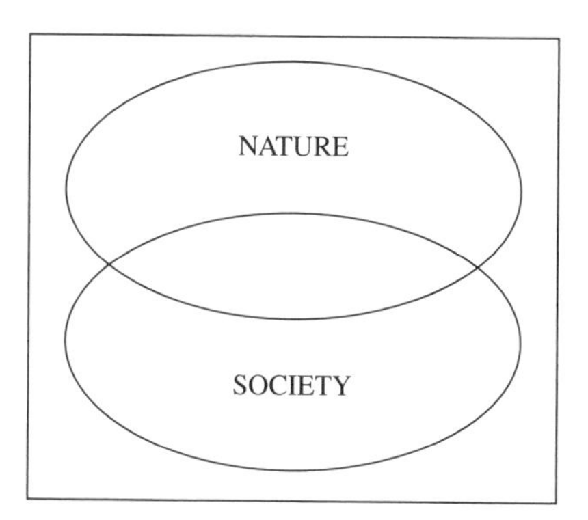

Figure 1. In Stegner's writing, the realms of the human and natural worlds are broadly overlapping but distinct.

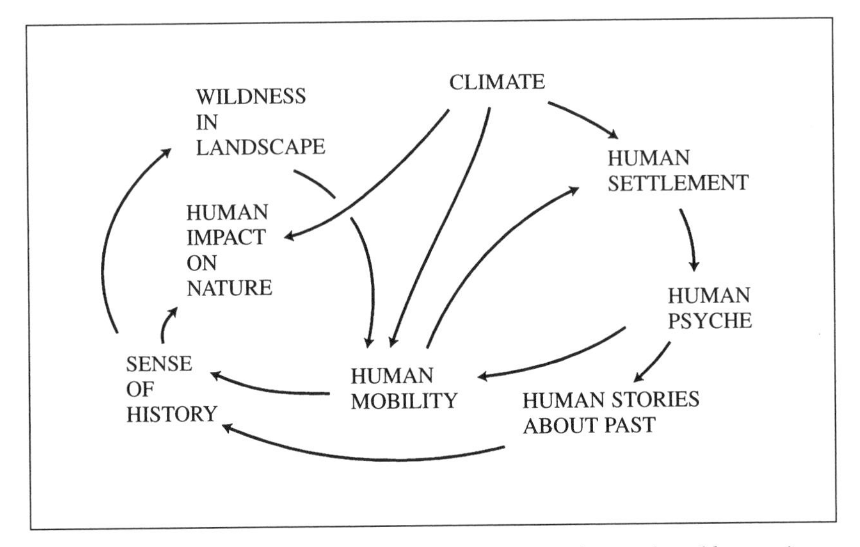

Figure 2. Various themes in the overlap of the human and natural worlds recur in Stegner's writing.

(Figure 3). These interactions, embedded throughout Stegner's work (but portrayed particularly effectively in *The Big Rock Candy Mountain*), create a flow of logical connections: the arid climate of the West forces a sparse human settlement, which, in turn, maintains the perception of a landscape of opportunity in which rugged individualism and independence are the expected personality responses:

> Aridity, and aridity alone, makes the various Wests one. The distinctive western plants and animals, the hard clarity . . . of the western air, the look and location of western towns, the empty spaces that separate them . . . the pervasive presence of the federal government as landowner . . . those are all consequences, and by no means all the consequences, of aridity. [pp. 8–9]

> Space, itself the product of incorrigible aridity . . . continues to suggest unrestricted freedom, unlimited opportunity . . . a continuing need for self-reliance and physical competence. [p. 111]

This human spirit of individuality encourages mobility—a periodic uprooting and pursuit of the big rock candy mountain, a spatial motion also

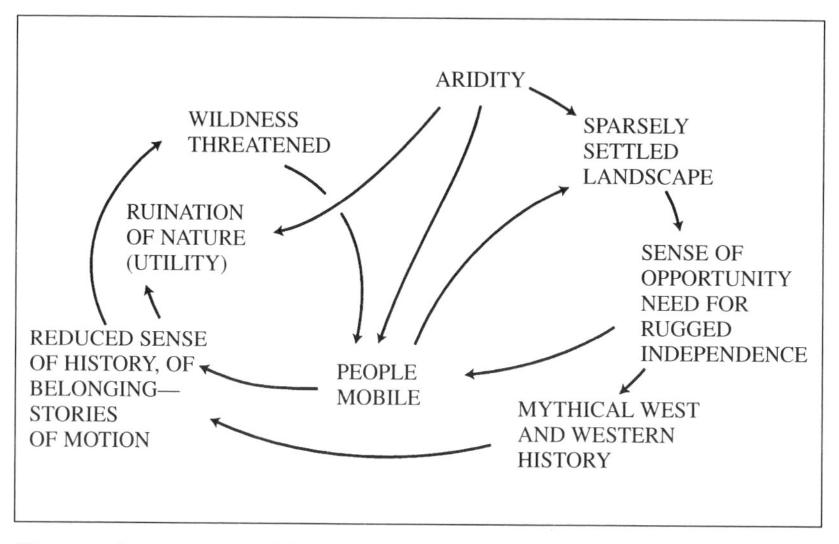

Figure 3. Stegner's view of the interactions between the human and natural worlds in the history of the American West is mostly a negative story. Primary sources: *American Places*, *Angle of Repose*, *Big Rock Candy Mountain*, *Bluebird*, *Living Dry*, *Wolf Willow*.

prompted by the recurring failures associated with a regional climate that offers too little water. This lack of sedentary lifestyles, moreover, perpetuates the pattern of sparse human settlement:

> As I've been reiterating ad nauseam for fifty years . . . aridity enforces space, which in turn enforces mobility. [p. 138]

> Mobility of every sort—physical, familial, social, corporate, occupational, religious, sexual—confirms and reinforces the illusion of independence. [p. 108]

The psychological propensities toward individualistic behavior also reinforce the mythical West of proud and independent—and sometimes noble—ranchers and miners, cowboys and fur trappers, all struggling against the wild landscape and the constraints of corrupting civilization:

> It was Crevecoeur's wild man, the borderer emancipated into total freedom . . . who really fired our imaginations, and still does. . . . What charms us in them is partly their daring, skill, and invulnera-

> bility, partly their chivalry; but not to be overlooked is their impatience with all restraint . . . and along with that competitive individualism and ruthlessness goes a rejection of any controlling past or tradition. [pp. 106 and 108]

Both the constant mobility and the inclinations toward defiant independence reduce the sense of historical belonging—to a society, to a group, to a place—and, instead, the stories that relate to the past are those of people moving: of motion.

> The West has never had a real literary outpouring . . . a lot of what has been written is a literature of motion, not of place. [p. 203]

> Migrants deprive themselves of the physical and spiritual bonds that develop within a place and a society. Our migratoriness . . . has left at least some of us with a kind of spiritual pellagra, a deficiency disease, a hungering for the ties of a rich and stable social order. Not only is the American home a launching pad, as Margaret Mead said; the American community, especially in the West, is an overnight camp. [p. 72]

As a consequence of such a worldview, people easily abuse nature as they modify it for utilitarian purposes—abuse facilitated by the fragility of the arid landscape—and this misuse prompts still more mobility. Apart from encouraging the abuse of nature for the sake of utility, this worldview has little use for the perpetuation of wildness in the natural landscape:

> We lived an idyll of miniature savagery, small humans against rodents. Experts in dispensing death, we knew to the slightest kick and reflex the gophers' ways of dying. . . . In the name of wheat we absolved ourselves of cruelty and callousness.[5]

> The vein of melancholy in the North American mind may be owing to many causes, but it is surely not weakened by the perception that the fulfillment of the American Dream means inevitably the death of the noble savagery and freedom of the wild.[6]

Taken together, this constellation of elements captures the most common themes of people/nature interactions in Stegner's writings. Comprising largely a negative perspective on these linkages, it looks to the past and sees a story of personal disillusionment, social coarseness, and environmental ruination. On the whole, however, this set of ideas cannot fulfill

Stegner's sense of place. It is historical, but negative, and thus lacks an ingredient critical to the concept of a sense of place: the ingredient of affection. It does not celebrate the West as a region.

A second constellation of people/nature interactions, again conforming to the general model, appears in Stegner's writings (Figure 4). Sometimes describing what has happened historically in a few certain locations in the West, this view looks mostly to the future. As in the first set of themes, here the elements of this one are linked together in a cohesive flow: the arid climate of the West results in a pattern of sparse human settlements, nucleated around the problems of persistence in a sometimes ungenerous environment. The resulting mutual cooperation and social responsibility demand a sedentary lifestyle in which adaptation to aridity is marked by restraint in expectations:

> In the actual desert, and especially among the Mormons, where intelligent leadership, community settlement, and the habit of cooperation and obedience were present, agricultural adaptation was swift.[7]

> The Spanish of New Mexico . . . are in other ways an exception. Settled at the end of the sixteenth century . . . New Mexico existed in

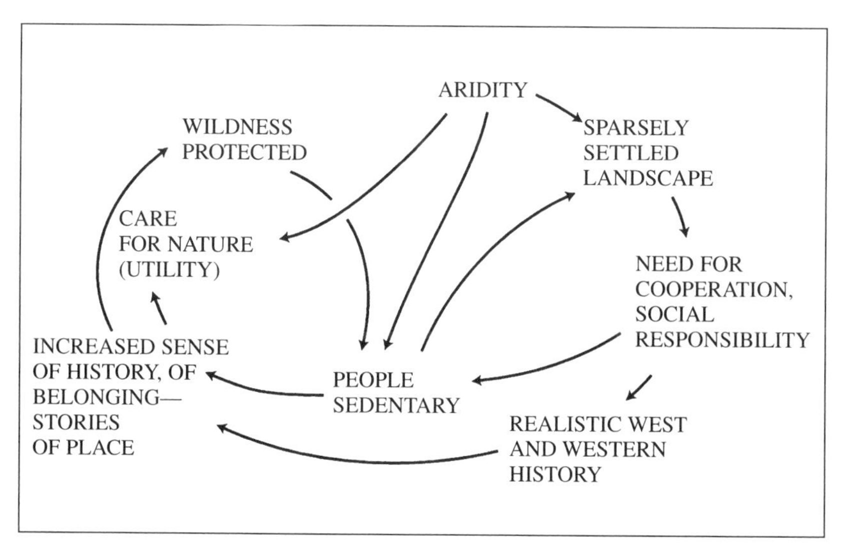

Figure 4. Stegner's view of the interactions between the human and natural worlds in a future American West, one with a stronger sense of place, presents a positive story. Primary sources: *Bluebird, Living Space, Wolf Willow.*

> isolation, dependent largely on itself . . . and during those two and a half centuries it had a high Indian culture close at hand to teach it how to live with the country. Culturally, the Spanish Southwest is an island, adapted in its own ways [p. 101]

In such societies, a more realistic perspective on what is needed to endure in the West fosters an honest assessment of the past. The spatially fixed settlement and the enlightened view of its genesis result in a sharper sense of history and a sense, too, of belonging to specific locales. Here the stories that are told are tales of place:

> Other phases [of Western settlement] have lasted by now for a century or more, and have formed the basis for a number of relatively stable communities with some of the attributes of place, some identity as subcultures of the prevailing postfrontier culture of America. [pp. 102–103]

> It is in places like these, and through individuals like these, that the West will realize itself, if it ever does: these towns and cities still close to the earth, intimate and interdependent in their shared community, shared optimism, and shared memory. [p. 116]

> The most quintessential West . . . [is] a little city as Missoula or Corvallis, some settlement that has managed against difficulty to make itself into a place. [p. 115]

In turn, the more conservative and enduring mindset promotes greater care in the utilitarian uses of nature—which also permits the sedentary lifestyle—and greater appreciation of the protection of wildness in the landscape:

> In the West it is impossible to be unconscious of or indifferent to space. . . . It . . . encourages, in some, an impassioned protectiveness: the battlegrounds of the environmental movement lie in the western public lands. [p. 112]

> [Wendell Berry says] that if you don't know where you are you don't know who you are. . . . He is talking about the knowledge of place that comes from . . . profound investment of labor and feeling. . . . No place, not even a wild place, is a place until it has had that human attention that at its highest reach we call poetry. [p. 205]

These positive interactions, or at least some of them, Stegner saw as descriptive of what actually happened in parts of the West. But more than interpreting the past, this optimistic set of interactions between people and nature projects what Stegner desired for the future of the West. And, because of this temporal perspective, the linkages in this optimistic set of themes are, as in the first set, unable to fulfill Stegner's sense of place: they express the affection but look, not to the past, but to the future. They portray a desirable long association of people in particular locales but find few examples where such associations have actually occurred. They describe not what has been but what might be.

These two—one that looks to the past, the other mostly to the future—describe Stegner's views of the interactions of people and nature in the West. But neither of them succeeds in wedding the definition of place with Stegner's own regional, place-identifying affection. What is missing?

In Search of Common Ground

At this point it is helpful, not to contrast the two sets, but to seek common ground between them. Not only does this mental exercise begin to construct a framework for Stegner's sense of place, but it also reveals Stegner as both an old-fashioned traditionalist and an avant-garde modernist (or even postmodernist). We can find examples of both traditional and modern traits in Stegner's work, and these reveal him as one who sought to blend perspectives.

Traditional Traits

Stegner was among our most celebrated contemporary writers, but, paradoxically, much of his work focused on what are generally regarded as traditional themes, approaches, and values. This orientation was evident not only in his writing, but in his conservation activities.

Environmental Determinism. In both sets of interactions, Stegner often portrayed the western environment as contributing to the development of certain human personalities:

> [The Great Plains] is a country to breed mystical people, egocentric people, perhaps poetic people. But not humble ones. At noon the total

> sun pours on your single head; at sunrise or sunset you throw a shadow a hundred yards long. It was not prairie dwellers who invented the indifferent universe or impotent man. Puny you may feel there, and vulnerable, but not unnoticed. This is a land to mark the sparrow's fall.[8]

Similarly, Stegner thought the natural world molded the American personality—that "the wilderness idea . . . helped form our character."[9] Even his admiration of John Wesley Powell—who envisioned a humane society derived from lifestyles consistent with Western aridity—and Walter Prescott Webb—who saw western cultural traits originating as responses to living in a dry land—suggests Stegner's fondness for explanations of human behavior and material culture that invoked deterministic links from the natural world.[10]

Such environmental determinism seems antiquated today, having long since passed from favor. Among most scholars, the mere suggestion of such reasoning is sufficient to brand a person as naive, archaic, or worse.[11] Yet the intellectual hostility toward environmental determinism may blind more than it enlightens. In their reaction against invoking the environment as a shaper of societal traits, some critics elevate other forces—notably the political system—as singularly deterministic,[12] thereby committing the same crime—oversimplification—as those who have championed the environment as causative. More moderate positions, which embrace aspects of environmental determinism, may see a blending of biophysical forces and humanistic commitments. Stephen Frenkel says this is true for the contemporary movement known as bioregionalism.[13] Essentially the debate over environmental determinism may represent still another expression of a dichotomy in human faiths—one driven by a conviction that both the earth and human cleverness are infinite, the other by a belief in limits, bounds, constraints. At the very least, the ease with which Stegner, obviously thoughtful in such matters, invoked the principle should make us pause and reconsider the unthinking and categorical rejection of all forms of environmental determinism.

Wilderness As an Initial Condition. Stegner's fundamental story line for the American past began with wilderness:

> [The person who lived on the American frontier] will know one thing about what it means to be an American, because he has known the

> raw continent. . . . Anyone who has lived on a frontier . . . has first renewed himself in Eden and then set about converting it into the lamentable modern world.[14]

Such a perspective is contrary to the modern view (represented in this volume by Elliott West), which emphasizes the fact that North America was indeed an inhabited place prior to European settlement, and that its native peoples humanized its landscape through myriad activities—agriculture, hunting, foraging, town building, landform construction, burning.[15] To invoke the imagery of a pristine wilderness, this critique argues, is to deny the basic humanity of its original inhabitants.

In many locales—the eastern side of the great midwestern prairies, for example—this vision of a humanized landscape in prehistoric times is reasonable and valid. But as a universal and unequivocal principle, it has itself become as misleading as the notion of a pristine landscape. In the American West, for example, the impacts of Native Americans on the region's natural systems may well have been modest: agriculture was restricted to certain rather small areas of the Southwest;[16] harvest of wild plants and animals, however important locally, likely declined in magnitude at ever larger scales, judging by regional vegetation surveys;[17] construction of both settlements and landforms humanized only specific locales, especially along the more populous Pacific coast, whereas elsewhere human numbers were low over most of the landscape;[18] fire regimes probably reflected vegetation flammability and weather—dry lightning and desiccated vegetation—more than Indian-set fires (although this issue remains contentious).[19] These suggestions remain speculative, but they are consistent with the character and functioning of western ecosystems.[20] Generally, then, Stegner was probably more right than wrong in seeing much, perhaps most, of western America of 1492 as a wilderness landscape primarily molded by natural forces.

Wilderness Preservation. Stegner did not simply endorse efforts to protect wild landscapes; he held such traditional efforts in the highest regard:

> We haven't really busted the whole continent. There are . . . a few million acres in wilderness that are permanently set aside. . . . There are also all the national parks, national monuments, and national historical sites, lakeshores, seashores, and recreation areas. There's quite a

> lot of . . . state land put aside in parks. Having really gutted the place in the nineteenth century, we've been doing much better in the twentieth.[21]

The strength of his conviction that wilderness deserves protection might even be described as hyperbolic. In his essay "It All Began with Conservation" (published in *Where the Bluebird Sings* as "A Capsule History of Conservation") he wrote: "If the national park idea is, as Lord Bryce suggested, the best idea America ever had, wilderness preservation is the highest refinement of that idea."[22]

This view that wilderness represents the "highest refinement" of nature protection is rejected today by many scholars, who see the wilderness ideal as too remote, too pure, too detached from human life. They argue instead for a vision of nature as everywhere, as inextricably intertwined with human activity and human history, as present in the common landscapes of everyday living. To appreciate this "second nature," the reasoning concludes, we must purge ourselves of the wilderness ideal.[23] Such a position, however, seems extreme and polarizing: Stegner himself valued both wilderness pure and nature diluted, suggesting that embracing one need not preclude embracing the other. To accept both, to see the two types of nature as desirable and compatible, itself expresses a romantic sensibility more consistent with the first half of this century than the second.[24] Today, by contrast, it is common to portray humans, simply and universally, as destroyers of nature. Again, Stegner seems traditional.

Modern Traits

Though he saw himself as a traditionalist, Stegner was hardly an anachronistic scholar out of touch with modern ideas.[25] In several ways, his views of the interconnectedness of people and nature reveal thoroughly contemporary perspectives.

Continuity Between Past and Present. Long ago Stegner observed that literary and historic portrayals of the American West tended to end without bringing the themes or peoples or events into the present—a temporal separation that contributed to the mythical character of the region and its history:

> I want . . . to underscore the point . . . about the absence of a present in western literature and in the whole tradition we call western. It re-

> mains rooted in the historic, the rural, the heroic. . . . That means that it has no future either, except to come closer and closer to the stereotypes of the mythic. . . . I think that you [should not] choose between the past and the present, you try to find the connections, you try to make the one serve the other. No western writer that I know has managed to do it, unless Wright Morris did it in his sequence of novels about Nebraska. . . . Maybe it isn't possible, but I wish someone would try. I might even try myself.[26]

Stegner, who wrote these comments in the middle 1960s, would eventually "try himself" when he employed literary devices and historic themes to link past and present in *Angle of Repose*.

Today, this temporal linkage is part of the agenda of the "New Western Historians," who have come to the fore over the past couple of decades.[27] Patricia Limerick has been especially outspoken in urging the articulation of such temporal continuity. She argues that the prominent issues of today's West extend backward unbroken into those of the nineteenth-century West: controversies over Indian resources and autonomy; border problems with Mexico; water allocation struggles; boom and bust cycles in extractive resources; immigration disputes; public land decisions. To understand the issues, either in the past or today, requires interpretation that is freed from single points of view—from "a preoccupation with the frontier . . . [and, instead, seeks meanings based on] the continuous sweep of Western American history."[28] Stressing "the connections" and avoiding "the stereotypes of the mythic"—these were Stegner's similar admonitions thirty years ago. The point here is not to claim that Stegner was "first," but to observe, simply, that his midcentury view of the need to tie past and present resonates favorably among contemporary historians.

Narratives or Stories. Stegner believed that the characterization of the past—or the present—was facilitated by the use of a particular literary form: the story or narrative. Much of his writing illustrates this belief (perhaps most obviously in *Wolf Willow*), but he also articulated the importance of storytelling as a means of creating the continuity between past and present that is a part of the sense of place:

> No place is a place until things that have happened in it are remembered in history, ballads, yarns, or monuments. Fictions serve as well as facts. . . . Occasionally we get loving place-oriented books such as

> Ivan Doig's *This House of Sky* and Norman Maclean's *A River Runs Through It*, but even while we applaud them we note that they are memorials to places that *used* to be, not celebrations of ongoing places.[29]

Today the narrative is being "rediscovered" as a literary style appropriate to the social sciences, particularly to the New Western History and environmental history.[30] From William Cronon's "Kennecott Journey" (1992) to Nancy Langston's *Forest Dreams, Forest Nightmares* (1995), from *Orion* magazine's "stories of the American land" to *Audubon* magazine's "Sense of Place," from Terry Tempest Williams' *Refuge* (1991) to William Kittredge's *Hole in the Sky* (1992), the narrative has become a fashionable style for writing about people and land in the American past.[31] Again, Stegner's admiration for the place-identifying effect of storytelling sounds entirely contemporary.

Place. Stegner was concerned with the characterization of place—with the development of a sense of place—even before he began to apply these labels to his efforts. In his later years he would come to stress the centrality of this development in the emergence of productive links between people and nature:

> History was part of the baggage we threw overboard when we launched ourselves into the New World. . . . Plunging into the future through a landscape that had no history, we did both the country and ourselves some harm. . . . Neither the country nor the society we built out of it can be healthy until we stop raiding and running, and learn to be quiet part of the time, and acquire the sense not of ownership but of belonging. . . . Only in the act of submission is the sense of place realized and a sustainable relationship between people and the earth established.[32]

Stegner's emphasis on the uniqueness of situations, the importance of particulars, the preeminence of historical circumstance, all of which are essential ingredients to the identification of place, are decidedly contemporary. In fact, in modern evaluations of the surface of the earth, the special, subjective, and personal character of place is much more fashionable—at least among those who study human phenomena—than the universal, objective, and functional analysis of space.[33] Again, Stegner's interest in place—including his 1981 book of narrative essays, *American Places*—marks him as fashionably modern.

The traits I have enumerated here are not the only characteristics of the two theme-sets. Nor do the traits define all the possible thoughts on people/environment linkages in Stegner's writings. One might wonder, for example, if he overestimated the importance of motion as a hindrance to the development of "placeness." After all, his own early life of almost constant motion did not preclude the emergence of such a sense in himself. Perhaps he overlooked the creative potential inherent in those who are particularly sensitive and receptive individuals. Moreover, he may have overlooked the possibilities of concerted effort to create such a sense in a human population. By contrast, Stegner may have underestimated the importance of formative years in people's lives—notably, certain periods of youth—in which attachment to place might emerge. Stegner's own life, for example, suggests the sensitivity of the preteen years.

Constructing Stegner's Sense of Place

The traits common to the two perspectives—the two models—define Stegner as both traditional and contemporary and help us to understand some of the detail in his reaction to the West. But the basis for his sense of place remains elusive. Perhaps it is time to turn from reductionist dissection and look instead in the other direction: at the initial definition of place. How might Stegner's interpretations of the links between people and nature differ from the constellation of elements—the meaning of place—with which this exploration began?

Stegner's view backward, which began with wilderness, stresses a critical role for the natural world—a much stronger role than is usual in articulations of the meaning of place. Nature molds human personality, provides an initial stage for the American experience, and deserves protection from the forces of modern society. But it is neither the pure, inviolate wilderness, nor the biodiversity so treasured by modern conservationists, that alone impressed him. His writings glow with nature in all forms in all places—winter rain on patio flagstones in California, desiccated soil or bold wind on the Great Plains, towhees scratching in the leaves and redtails soaring overhead south of San Francisco, minks cavorting in the Uintas, the "whale-like shapes" of Utah's high plateaus,[34] the "exciting salt-flat odor" of Great Salt Lake,[35] the home-bringing scent of wolf willow on the

Frenchman River. It was the natural world in all its expressions that fostered development of an intensified sense of place—particularly in the American West.[36] Clearly, for Stegner, places were not places, imbued with affection, unless they radiated with nature.

Stegner's view forward in time, seeing the wildness of the West as promising for a wiser and gentler human society, suggests still another contrast with the conventional definition of place. In fact, the very notion that places assume meaning, take on a sense, as a consequence of possible futures is contrary to the usual emphasis on familiarity based on long historical association. Stegner's thinking offers a time-transcendent quality: past, present, and future linked in stories that define place.

Stegner's vision, moreover, was that of the utmost optimist: "Give it a thousand years," he wrote of the possibility that Whitemud—the isolated Great Plains settlement of his youth—might eventually become something cultured and civilized.[37] His patient optimism also appears metaphorically with Oliver Ward's careful cultivation of roses in *Angle of Repose*:

> Grandfather made a rose. He made a dozen roses, in fact, trying for just the right one. You know how long it takes to cross and fix a hybrid rose? Two or three years. He could never get just what he wanted. . . . If he got one that would go on blooming, something was wrong with the color, or it didn't have any odor. If he hung onto the color he wanted, it bloomed in May and was done for the season. Eventually he had to give up and accept a brief, early blooming.[38]

Although in interviews he expressed pessimism about the future of the West,[39] his stories retain a sense of the possible, of hope, of progress, of the future. Stegner's affection for the West lay partly in the prospects for what the place might become.

To observe that Stegner enlarged the meaning of the concept of place—stressing nature, looking forward—does not mean that he neglected either of the contrary characteristics, civilization or history. In fact, in a conceptually larger view, perhaps Stegner's meaning of place was unusual, and uncommonly felt, precisely because he embraced what are usually taken to be contrasting poles: nature and nurture, wilderness and civilization, nature preserved and nature utilized, past and present, stories of what was and what might be, the high culture of poets and the wildness of mountain water. One of his virtues was his lack of blind commitment to either one extreme or the other in what are often polarized ideologies. The wide arms of his embrace, moreover, did not reduce the differences in the polar characteristics to some bland yuppie hug. Like Susan and Oliver Ward, who, in spite of tragedy, maintained their individual upright stances and yet achieved a linking angle of repose, Stegner's sense of place, ultimately, accomplishes the repose of wholeness.

Stegner's Lesson

> When Bruce drove west in June . . . he drove directly from rainy spring into deep summer, from prison into freedom. . . . He was a westerner, whatever that was. . . . As long as the road ran west he didn't want to stop, because that was where he was going, west beyond the Dakotas toward home.[40]

It was not only a landscape through which, and to which, Bruce drove. It was also a place—a place rich in natural endowment and in human potential. His trip carried Bruce through space, propelled him through time. It was a trip literally across the plains but also metaphorically across the human mind. May all of us, alone or as a society, like Bruce, move forward always conscious of the past—because the route behind helps to determine the trajectory of motion. In the gathering dusk, the headlamps of the car illuminate the rich sunflowers blooming on the road shoulder and the last redwings wheeling into the swaying grasses, but they bring our eyes back to the pavement before us, focused by the centralizing light of Wallace Stegner's sense of place, moving ahead toward all that is "gentle and civilized," moving ahead from the "unfinished beginning," moving ahead, "west beyond the Dakotas toward home."

CHAPTER TWELVE

FIELD REPORT FROM THE NEW AMERICAN WEST

RICHARD L. KNIGHT

IN HIS ESSAY "The Sense of Place," Wallace Stegner wrote:

> The frontiers have been explored and crossed. It is probably time we settled down. It is probably time we looked around us instead of looking ahead. We have no business, any longer, in being impatient with history. We need to know our history in much greater depth. . . . Plunging into the future through a landscape that had no history, we did both the country and ourselves some harm along with some good. Neither the country nor the society we built out of it can be healthy until we stop raiding and running, and learn to be quiet part of the time, and acquire the sense not of ownership but of belonging.[1]

A sense of place is an underlying theme in much of what Stegner wrote, whether fiction, history, or autobiography. This fascination with home and place might have been the result of how he was raised, for Stegner was the son of a boomer, a man who was the epitome of a raider always looking for the "main chance." Stegner was, in his own words, "born on wheels."[2]

Considering that Stegner was literally a son of rootlessness, it is perhaps surprising that he captured for many the yearning sense of home in the West. Perhaps his very homelessness made Stegner alert to the concept of home and belonging. Perhaps his lack of a residence caused him to imbue his writings with a feeling of loneliness and the need to belong somewhere.

In his essay "Finding the Place: A Migrant Childhood," he writes, "Growing up culturally undernourished, I hunted the Big Rock Candy Mountain as hungrily as ever my father did, but his was a mountain of another kind. He wanted to make a killing and end up on Easy Street. I wanted to hunt up and rejoin the civilization I had been deprived of."[3]

Critics have acclaimed Stegner, much as they did Bernard De Voto before him, as the chronicler of a vast region known variously as the Mountain West, Intermountain West, or the Big Open. Today there are legions of people who credit Stegner for helping them develop an understanding of the area defined by Mountain Standard Time, much as others pay homage to William Faulkner for helping explain the complexities of southern culture. Of course, Stegner's writings embraced not only the human and natural histories of the American West: his vision loomed larger, covering the Midwest, New England, and other landscapes and cultures. Indeed, if all one had read of Wallace Stegner was *Crossing to Safety*, one would deem him an exemplary writer of New England. In these regional landscapes that fell outside the American West, he wrote also of home and belonging. Who can forget the pang of regret Stegner expressed for the New England landscape losing its family farms in his short stories "The Sweetness of the Twisted Apples" or "The Berry Patch?" His evocative writing captured the pathos of families that tried to live on the land but eventually succumbed to the realities of economic and social upheaval:

> In the open sunlight she sat on a gravestone and thought how it might have been to be the last family left on such a road, and to bury your child among the dead who had been gathering for a hundred and fifty years, and then to move away and leave the road empty behind you. She imagined how it might have seemed to some old grandmother who had lived in the village for eighty years, watching the hill farms go dead like lights going out, watching the decay spread inward from the remote farms to the near ones, to the place next door.[4]

But through it all, Stegner left a literary trail that rebelled against America's love affair with motion and pleaded the need for place and belonging.

The New West

Today, Stegner's writings about becoming a "placed person" and the need to stop the raiding are more relevant than ever, for once more the American West is booming. Not since the land-rush days has the region seen so much in-migration. Our intermountain states are growing at annual rates that exceed those of many developing countries: Nevada 3.9 percent, Idaho 3.1 percent, Colorado 2.9 percent, Utah and Arizona 2.7 percent, New Mexico 2.2 percent.[5] Even Wyoming, a state that has bled people for decades, shows a net annual population increase of 1.2 percent. The Mountain West is not only the fastest growing region in America; its growth rate rivals that of Africa and exceeds that of Mexico. Amid the heart-stopping grandeur of the Colorado Plateau, the resident population doubled between 1960 and 1990—and grew by another 15 percent between 1990 and 1994. La Plata County in Colorado showed a 14 percent increase in population between 1990 and 1994.[6] And so it goes, county by county, valley by valley. The West is filling up, once more the destination point of dreamers, boosters, and raiders, although this growth is unlike any the region has seen before.

The migrants in this latest land rush are an interesting group. They are not the landless poor wishing to homestead on the raw western prairies of the late 1800s. And they are not the miners, oil field roustabouts, or gas field workers descending upon sparsely populated western towns to ride the latest energy boom. No, these are America's finest. They are the modem movers of cyberspace who can live where they choose. They are often the affluent who can afford a trophy home in the Rockies. They are those who heretofore have been denizens of our metropolitan centers—those who, as the saying goes, can sell their bungalow in the Bay Area and buy a ranch in New Mexico. Called variously "cappuccino cowboys," "modem movers," or "lone eagles," they were profiled in *Time* magazine this way:

> In the Rocky Mountain region, it's not taxi drivers anymore—it's professional people who realize they can locate anywhere and live by their wits. Many were middle managers who were forced off the corporate gravy train in the latest recession and downsizing and said, "Why live in New York or L.A.? I can have a modem and a fax and live anywhere I like."[7]

What is at the root of this latest surge of restlessness? What would cause a family to uproot and head toward America's hinterland? After all, this was a region long passed over. In 1950, the eight Mountain West states, which comprise approximately one-third of the landmass of the lower forty-eight states, contained just 3 percent of the country's population. Indeed, in each decade after 1900 through the 1950s the region's population grew at a slower rate than in the previous decade.[8] While the hinterland was passed over by Americans redistributing themselves, the Pacific Rim states—California, Oregon, and Washington—boomed.

What kept the Mountain West from becoming the "American backwoods" that was predicted for it?[9] Why did growth in California slow and that in the Mountain West skyrocket? This growth is being fueled by the "floating baseline," a concept that captures the imbalance in indicators of quality of life between two different areas. People flock to Fort Collins, Colorado, where I live, because crime, traffic congestion, air pollution, and the cost of living are less than where they came from, be it Des Moines or Pasadena. Let John Hough, as reported in *Time*, describe his reasons for picking the Intermountain West:

> Comin' round one side of the mountains is John Hough. On New Year's Day, the 43-year-old police sergeant, a veteran of the Los Angeles riots, took in a view of California's San Bernardino Valley—as best he could. A blanket of smog had smothered the landscape. "Look at that crappy air," he said to his wife Patricia, 32, as they drove home from a Colorado vacation. "Why are we spending the young years of our life in California when we like Colorado so much?" In the next three months, Hough would turn in his badge and trade his rented Orange County, California, condo for a $103,000 cedar house on 2.5 acres of woodland in idyllic Bailey, Colorado. "It's been tough looking for a new job," he says, gazing at snowcapped Mount Evans through the tall pines outside his picture window. "But we have no regrets. It's been a great move—for family, for affordability, for all-round quality of life.[10]

This imbalance in quality of life between the towns in the Mountain West and areas outside the region suggests people will continue to move to the West for decades to come. California, a state of over 32 million people, has almost twice the population of all the Intermountain West states combined and stands as a rich source population. Between 1992 and 1993, for example, over 155,000 Californians moved to the inland western states.

This growth is not occurring evenly across the West. Not unexpectedly, considerable growth is still taking place where growth has already occurred. Along Colorado's Front Range, for example, almost 80 percent of the state's population is planted along the interstate highway from Fort Collins to Pueblo, with Denver, Queen City of the West, at its epicenter. Not surprisingly Aspen and Telluride, Santa Fe and Taos, Sun Valley and Jackson

Hole, Bozeman and Park City, continue to boom. Mountains have always attracted émigrés and still are powerful magnets. In Montana, for example, counties along the Continental Divide grew an average of 10 percent during the past decade, while those not touching the divide lost population.[11] Combined, the twenty counties of the Greater Yellowstone Ecosystem have grown at a rate 33 percent faster than Montana, Wyoming, and Idaho as a whole.[12]

But the growth is occurring elsewhere—in what until a short time ago had been working landscapes, where people dressed in canvas and denim bearing the brand names Carhardtt and Key. Buffalo, Pocatello, Silver City, Grand Junction, and Springerville are all experiencing growth like they have not seen since the days of whatever mineral, logging, or energy boom defined the town's past. What is so unexpected about this growth is where it is occurring. When farmers and homesteaders failed on the land—beginning in the early 1900s and continuing as America changed from an agrarian to an urban society—people moved from the country to the cities. Today people in the West are moving from the cities to the country. This is more than the suburban sprawl that rims the cities of the West and characterizes virtually every other urban zone in America outside the Buffalo Commons. Following the century-long redistribution of people from farms to the cities, a reversal began in the 1950s as people left the cities and moved into suburbs. In the 1980s, regional metropolitan areas in general began growing at faster rates than nonmetropolitan areas. The Mountain West, however, was the exception. There the nonurban counties increased faster than the nation as a whole, and more quickly than regional metropolitan areas.[13]

In the American West, land that is not public is private. And what is not in cities and towns is likely to be farms and ranchland. Therefore, most of this growth is occurring on what were once farms and ranches. Since 1978, for example, Colorado farmland has declined by ninety thousand acres per year.[14] Between 1969 and 1987, in Park County, Wyoming, 19 percent of the farmland was platted for subdivision; in Teton County, Idaho, the rate was 16 percent; in Gallatin County, Montana, the rate was 23 percent.[15] The New West is a place of ranchette developments—rural subdivisions as vast as the former ranches they now occupy (Figure 1). We are creating a new landscape: evenly divided into twenty- to forty-acre parcels laying across the land like a gigantic grid, as if a giant prison door had fallen flat upon the earth. What are the implications of altering forever these

Figure 1. Mailboxes of ranchettes on what was once a working cattle ranch in Colorado.

working landscapes and saying goodbye to family ranches? Stegner described such places in his vibrant and evocative essay "Crow Country":

> Right below us, sunken, tight, miniature in big country, the West Rosebud kinks and races through its belt of cottonwoods and willows. Strung along it for a mile or so is the Bench Ranch. Its house, barns, and corrals, its several old homestead buildings, its hayfields with their stacked bales, its red angus cattle as solid as boulders against the opposite slope, make up a picture that is intimate and humanized. It means neither exposure, nor longing, nor the impersonal protection of the mountains, but people, sanctuary, something that men have made for their safety and comfort in an indifferent universe.[16]

And what a presence this new wave of emigrants brings to the land! The topography of western lands is being reshaped as new homes reconfigure the natural contours of ridgelines and peaks (Figure 2). Indeed, many of these homes have unintentionally desecrated views that older residents

Figure 2. Home built along a scenic skyline in Colorado.

have long cherished, whether as their mountain backdrop, a canyon rim, or a panoramic vista. Early settlers and ranchers, by contrast, dug in, found the sheltered swales, located the sites in coulees and against cliffs, carved sod huts out of hillsides—wherever they could find shelter from the winds, the snow, the cold.

Today homes being built in the New West are quite different: they are bigger (Figure 3). Indeed, they are sometimes so big that even the gilders feel guilty and attempt to encourage yet newer settlers to show moderation. Pitkin County, Colorado, recently adopted a limit on house size: 15,000 square feet and no more than 4,750 additional square feet in basement and garage.[17] In and around the resort town of Jackson, Wyoming, concern has been increasing in response to new homes as large as 23,000 square feet. Recently, Teton County adopted a comprehensive plan that would limit single-family homes to 8,000 square feet of "habitable space."[18] These monuments to wealth and a new American dream are all the more inexplicable as contemporary families are considerably smaller than those of the pioneers.

Figure 3. A typical "trophy home" on what was once ranchland in Colorado.

The Costs of Life in the New West

What are the ecological implications of this new settlement pattern? Although approximately half of the West is publicly owned, much of the remaining private land is highly productive. Under the prevailing patterns of European settlement, riparian areas were homesteaded first and therefore released from the public domain. Prime development property includes wetlands, streams, and rivers. When these lands were part of farms and ranches, the population was scattered thinly along these riparian areas. Today, however, the distance between homes is measured in yards rather than miles. Ranches adjacent to public lands are eagerly bought and subdivided as well. Suddenly national parks, national forests, and BLM lands are encumbered by as many as eight layers of ranchette development (Figure 4).

This New West is quite different from the past landscape in that people are now congregating in rural areas and adjacent to our public lands. This translates into more dogs and cats, more automobiles and road-killed wildlife, more landscaping with nonnative plants, more halogen lights at

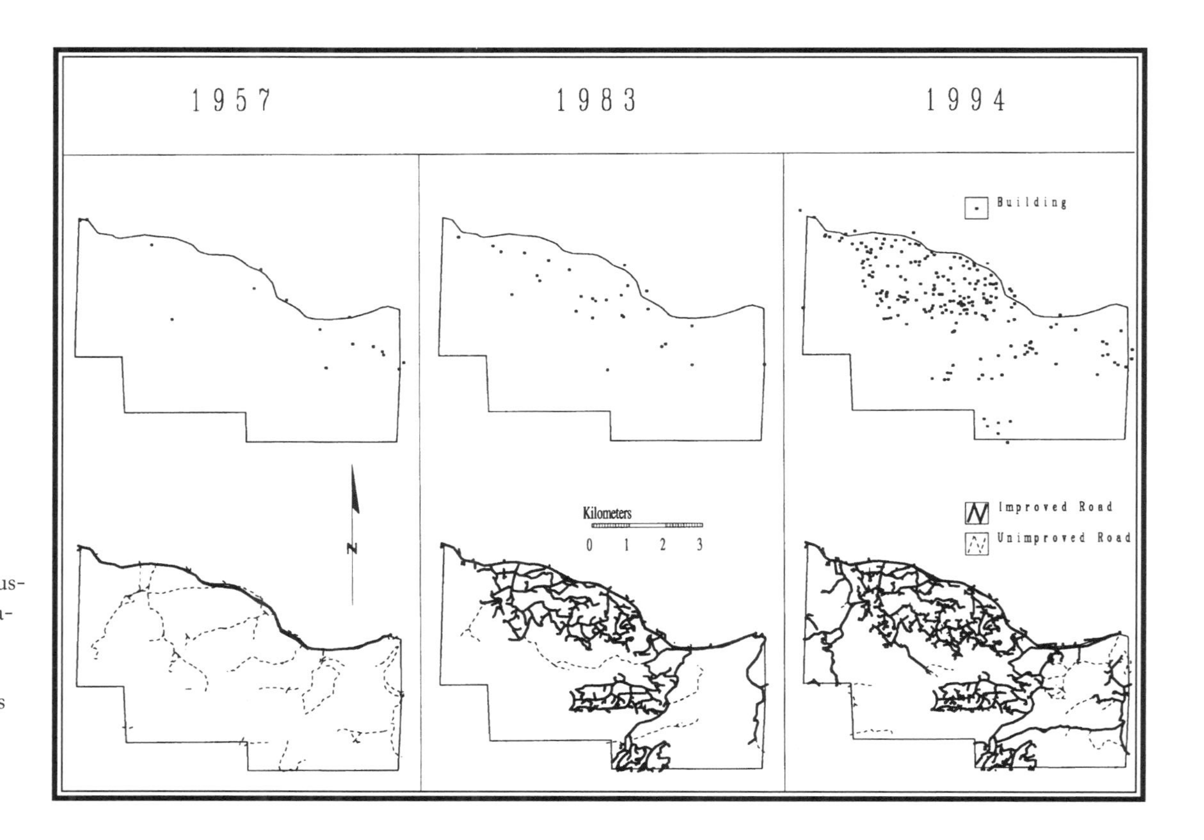

Figure 4. Changes in numbers of ranchettes and roads after a ranch was sold. This rural housing development is adjacent to the Arapaho-Roosevelt National Forest, Colorado, and is representative of many across the American West.

night, more noise, more people walking across the land. Preliminary studies indicate that this invasion will result in an altered natural heritage, encouraging generalist and human-adapted species and discouraging species sensitive to human activity.[19]

Such changes, of course, have not been confined to the Mountain West. In the hills above Palo Alto, California, Wallace and Mary Stegner lived from when they "could see not a single light at night except stars and moon" to the day when "the bulldozers are at work tearing up the hills for new houses—twelve here, thirty-one there, fifteen across the gully."[20] Because he lived in what became a rural subdivision, and because he was a keen observer, Stegner wrote compellingly about the loss of wildlife that comes with increasing human numbers:

> We live among the remnants. Feral housecats, as efficient at hunting as the foxes . . . and more efficient at breeding . . . have almost cleaned out the gophers, and I suppose the mice as well. That means that most of the animals who depended on that food supply couldn't make a living now even if we restocked the area with them. Even if there were enough food, the other hazards would be fatal. The last pair of great horned owls went several years ago, both of them apparently indirect victims of the poisoned carrots that some neighbor put out for the gophers. Both came to die in our yard. As one of them squatted in the patio, panting with open beak in the full sun, I passed my hand across above him to cut off the sun from his blinded eyes, and watched the slit of pupil widen instantly in the shadow, a miraculously controlled lens shutter, before the fierce yellow eyes went out. It wasn't quite what Aldo Leopold saw as the green fire went out of the eyes of a dying wolf, but it carried something like the same message, and it shook me.[21]

What changes do these rural developments portend for stewards of our nation's resources? There will be pronounced changes both within and beyond our parks and forests. From within, we will see more outdoor recreation and less commodity extraction through logging, livestock grazing, and other uses. After all, these new people move west for amenity values—for wildlife, recreation, and scenery—and not to participate in logging, mining, and ranching cultures. Heretofore, we have viewed outdoor recreational impacts on wildlife as benign. Increasingly we are beginning to appreciate that this is not the case. Outdoor recreation, if poorly managed, can be quite as detrimental to wildlife as logging and mining, although in less dramatic ways to the landscape.

This raises a whole new suite of problems for public land managers along their boundaries. When suddenly you have neighbors rimming your boundary who view fire as a legitimate threat to their homes, rather than an essential ecological process, can fire still be used as a management tool? When suddenly the public can enter national parks across the fence rather than through the park entrance, what will be the effects of the ranchette owner's chainsaws, dogs, cats, weapons, garbage, and exotic plants? When public lands are seen as the source of black bears, mountain lions, and ungulates, all of them suspect as threats or nuisances, does wildlife win? In Colorado, black bears now get "two strikes and they are out." In 1995, more than forty bears met their demise due to their attraction to garbage at ranchette developments.

I suspect that our managers of parks, forests, and other public lands will look back longingly to the days when their neighbors were simply ranchers. Stegner feared the impact of these hordes of new neighbors shoved up against the boundaries of our public lands: "The worst thing that can happen to any piece of land, short of coming under the control of an unscrupulous professional developer, is to be opened to the unmanaged public."[22] When people enter the public lands from their doorsteps rather than through designated entrances, how can managers manage?

What are the economic implications of this new pattern of development? New subdivisions are springing up so fast in Gallatin County, Montana, that the road department cannot keep pace in assigning new addresses and people are making up their own—sometimes creating duplicate addresses and adding confusion for firefighters and sheriffs. Gallatin County has seen 108,425 acres of land subdivided without any review since 1973.[23] This means that 92 percent of the county's subdivisions were completed without any consideration of their impact on roads, schools, electrical service, telephone service, public gas or private propane service, public water or private wells, public sewer or private septic systems. These properties are farther from towns and will yield more vehicle miles driven daily than city subdivisions, thus adding to air pollution and traffic safety problems.

Do the property taxes that come off these homes cover the costs to the county? Or are the economic costs of rural sprawl just as hard on the taxpayer as on Mother Nature? A study in the Greater Yellowstone region has found that for every dollar raised in revenue from residential property, the county government and school districts had to spend $1.45 to provide services there. Conversely, only $0.25 had to be spent to provide these services to agricultural land and open space.[24] Are these inhabitants of rural properties willing to pay taxes that compensate government for services? In Larimer County, Colorado, the county sheriff recently requested additional funds for officers to patrol his district with its far-flung rural ranchette developments. The voters said no. What about the children in these distant subdivisions? Is the one-hour bus drive to and from the city schools worth the health and happiness the family derives from living in the country?

Finally, as we busy ourselves with remaking the western landscape, what social costs do we gather as we convert a working landscape, ranch by ranch, valley by valley, into rural subdivisions and second homes of absentee owners? Stegner, in considering the future of the western ranch landscape, asked:

> Is that the future? We want to know. This is obviously country where people and the land get on well together, where the prevailing economy not only doesn't harm the land, but with good management may sometimes improve it. The life here is rich, strenuous, and satisfying. Can that condition hold? Does the family ranch get absorbed into the corporate ranch owned by a bank or by a limited partnership whose primary goal is tax writeoffs? And will the corporate ranch stick to cattle, or will it turn to more profitable uses for the land—dudes, summer cottages, subdivisions, vacation condominiums? Can country like this remain a difficult but satisfying workground that enlists body, brain, and heart, or is it inevitably going to get turned into a playground, with every deterioration that becoming a playground connotes? What is to prevent entrepreneurs from coming here, as they have come elsewhere, and building a Big Sky or a Ski Yellowstone or a Sun Valley? Is cattle-raising the highest economic use for this land, or is it an anachronism and an indulgence? Can the family with a few hundred acres and a hundred head of cows still make it?[25]

This New West of rural subdivisions replacing ranches, seemingly overnight, sometimes keeping only the name of the ranch on the housing development entrance sign, will be quite a different place. In some respects the American ranching culture shares similarities with our rich Native American cultures; while the latter certainly qualify as indigenous, the former is approaching this status. When family farms and ranches can count back six generations of kin living and working a piece of land, does this give them any special considerations in our helter-skelter rush to remake the West? I'm not sure. But one thing is certain: this New West will seem a more cold, impersonal, and lonely place as Volvos and Saabs carrying busy people with cellular phones replace old pickup trucks with people going about the chores of living and working the land.

Stegner too saw similarities between ranching families and Native Americans. In one of his most heartrending and evocative essays, "Crow County," he spoke about ranchers and Indians:

> And observe a cosmic irony. Out on the plains, the tamer country onto which the Crows were forced in the 1880s turns out to contain six billion tons of strippable low-sulphur coal. An equal amount lies under the grass of the Northern Cheyenne reservation next door. . . . The modern Crows can grow rich, if they choose to adopt white styles of exploitation and destroy their traditional way of life and forget their mystical reverence for the earth. Meanwhile the whites who now live in the heart of the old Crow country, as well as many who own or lease range within the present reservation, fight against the strip mines and power plants of the energy boom, and in the face of rising land costs, high money costs, high machinery costs, high labor costs, and uncertain beef prices work their heads off to remain pastoral. . . . There is a true union of interest here, but it is also a union of feeling: ranchers and Indians cherish land, miners and energy companies tear it up and shove it around and leave it dead behind them.[26]

As Stegner pondered what *High Country News* called "cultural genocide,"[27] he reflected on the costs the rural West might pay as people from outside the region moved in:

> There is both good news and bad in a rural migration, a mixed blessing as mixed for the newcomer hunting tranquility and self-determination as for the townsman whose community is invaded. Those who come in from outside with the specific intention of exploiting an opportunity to make a profit are bad news for both. But at least part of the bad news for the natives is a change in the status quo so rapid that it can't be comprehended, much less resisted. . . . Newcomers change their location, and even some aspects of the way they live. They sel-

> dom change their ethics, morals, cultural values, politics, or social habits. To a conservative native, a Johnny-come-lately urban egalitarian with his talk of zoning, planning, and pollution controls is hard to take. . . . On the other hand, newcomers congregating in any numbers begin to reproduce the problems they fled from in the first place. They may bring in some services that the community never had and doesn't want. Or they become the growth, expansion, and development that they themselves deplore. Their children need schools, their sewage has to be dealt with, their roads need plowing, their lives and property need protecting. They bid up the price of land and houses, and taxes go up and up. Oldtimers can't afford to keep their own places. Somebody gets the idea of attracting tourists. Craftspeople follow the tourists, and more tourists follow the craftspeople.[28]

What is the future of this New West, attracting these hundreds of thousands of new people living in the country but working in cities or striking paydirt in cyberspace? They live on a piece of land, but do they understand it? Are they here to simply ranch the view, enjoy the amenities of the Mountain West, and thank God they no longer live where they left? Because ranchette developments thirty miles from the nearest city are not part of a town, to what community do they belong? Because they are not dependent on the soil for their living, can they find a connectedness to something from which to build communities? Will they slow down, stay at home, and learn the human and natural histories of where they live? Will they come to understand how their dogs and cats, their exotic plants, their halogen lights beaming through the once black night, are altering the wildlife and silence and peace they thought they were heading for? Will their kids grow up, go off to school, and move back to the ranchette to work? Will they come to learn how to graze their horses to minimize overgrazing, how to place their access roads to minimize soil erosion, learn to live with rattlesnakes, black bears, and mountain lions? To appreciate the sublime beauty of their own landscape and refuse to build on its ridges and cliffines or up against the immediate streambank? Do these ranchettes promise more than a crowded and congested Mountain West? Stegner thought they might, though he uttered a caution:

> Rural America may well be where much of what has been lost during the last few decades gets reborn or rediscovered—things like family, community, occupational integrity, moral and social responsibility, a measure of self-sufficiency. The culture of narcissism finds it sterile

> soil. It is too quiet for the "me" generation. It offers more natural than manufactured diversion. But like so many American habitats, the small town is fragile. If its immigrants do not come to it with a degree of humility, with an understanding not only of its human ecology but of Paradise Lost as well, they can destroy it before it can gather itself and resist.[29]

How can we continue to populate the Intermountain West without spoiling it—without paying the economic, ecological, and social costs that continue to mount? There is much we can do. For starters we must honestly address the challenge that Wallace Stegner posed when he wrote:

> There are ways for people to exist in some measure of harmony with their natural surroundings, even though you might rather there be no people at all. But these ways require that we exhibit forethought and planning and aesthetic sensitivity, and they also require, particularly in this day and age, that we abandon the notion that a property owner can do anything he please with his property simply because he owns it and because that's the way it has always been. On occasion communities have managed to figure this out before it is too late.[30]

A New Sense of Hope in the New West

Despite the litany of complaints I have lodged against our New West, there is no justification for despair. It is not yet the time to capitulate and look for that last remaining place to flee to. Wallace Stegner, like all who struggle with challenges of the heart, vacillated between hope and despair. But he gave us reason for optimism when he wrote:

> There are signs of a change in American expectations, an alteration of the free land and unregulated individualism myths still clung to so desperately by the sagebrush rebels and "new Federalists." . . . The environmental movement in this country is at least as old as its first spokesman, William Bartram, but only in the last fifteen to twenty years has it been joined by millions of Americans who have learned a new respect for the land and acquired a new consciousness of their relationship to it. At long last it seems that ordinary citizens have become less commonly raiders and more commonly conservers and stewards of the only continent they are ever going to possess.[31]

His words are true. For every assault on this region, there are counterattacks by people living there who have come to appreciate the power of place and belonging. Historian Patricia Limerick has written: "If Hollywood wanted to capture the emotional center of Western history, its movie would be about real estate."[32] In response to the gluttony of land speculation and development, western communities are beginning to insist on what Colorado's Governor Roy Romer calls "smart growth." Acknowledging that the passes leading into the Mountain West cannot be barred, citizens are instead demanding that planning and cooperation be keywords in development. Meeting stiff resistance from those who view the West as the land of I-can-do-whatever-I-want, people are insisting that our leaders operate out of an environmental ethic that includes healthy communities and healthy ecosystems.

Increasingly our western communities are agreeing that a farming and ranching culture is worth preserving—that there are many values associated with families who can live on the land, provide open space, and protect biological diversity at no cost while producing food for an expanding population. Parts of the West still remain in ranching and farming. These endeavors should be acknowledged by the New West as essential to the region's cultural fabric. Tourists who linger in the recreational meccas of western valleys and parks agree. Economists asked visitors to Steamboat Springs, Colorado, if their decision to visit Steamboat in the future would be influenced by continued conversion of ranchland to other uses, such as golf courses and resorts. Just under half responded that they would seek other resort areas and not return to Steamboat. An overwhelming majority of visitors said the unique natural and man-made assets of ranchlands contributed to their enjoyment of their vacation.[33]

Moreover, ranchers and environmentalists alike are doing much to maintain ranching as a viable economy. Organizations as diverse as the Colorado Cattlemen's Association, The Nature Conservancy, and the Greater Yellowstone Coalition are exploring ways to maintain a ranching economy. Along another front, people are forming land trusts to protect open space and curtail development close to home. Only a few years ago Colorado had but a few land trusts. Today there are thirty-three, and more are forming each year. And they have a treasure chest from which to draw. Colorado's citizens decided to dedicate a portion of the state lottery proceeds to protect wildlife habitat and open space. In its first three years, this program provided $30 million for worthwhile projects.

These private/public partnerships have expanded to city and county open space initiatives. Increasingly across the Mountain West neighborhoods are agreeing to tax themselves to protect areas that define the community—a ranch on the edge of town, an intact riparian area, a remnant of short-grass prairie. The City of Boulder and Boulder County have protected almost 100,000 acres of open space dedicated to agriculture, wildlife habitat, and recreation.

Our public land management agencies are also doing their part. After a near-century of conflict over public land management, agencies from the U.S. Forest Service to the Department of Defense have equilibrated their gyroscopes and rediscovered Aldo Leopold's land ethic. Today our public lands are increasingly being stewarded under the concept of "ecosystem management," which acknowledges that our lands should not only provide commodity and amenity values but should also conserve native biological diversity and natural ecological processes.[34] Ecosystem management places primary emphasis on land health; commodities and amenities can be provided only to the degree that these uses are sustainable and do not degrade the land, its components, or its functions. Wallace Stegner clearly understood what ecosystem management was before the agencies reinvented it. He gave us one of its clearest definitions:

> The protection by these various agencies is of course imperfect. Every reserve is an island, and its boundaries are leaky. Nevertheless this is the best protection we have, and not to be disparaged. All Americans, but especially Westerners whose backyard is at stake, need to ask themselves whose bureaus these should be. Half of the West is in their hands. Do they exist to provide bargain-basement grass to favored stockmen whose grazing privileges have become all but hereditary, assumed and bought and sold along with the title to the home spread? Are they hired exterminators of wildlife? Is it their function to negotiate loss-leader coal leases with energy conglomerates, and to sell timber below cost to Louisiana Pacific? Or should they be serving the much larger public whose outdoor recreations of backpacking, camping, fishing, hunting, river running, mountain climbing, hang-gliding, and, God help us, dirt biking, are incompatible with clear-cut forests and overgrazed, poison-baited, and strip-mined grasslands? Or is there a still higher duty—to maintain the health and beauty of the lands they manage, protecting from everybody . . . the watershed and

> spawning streams, forests and grasslands, geological and scenic splendors, historical and archaeological remains, air and water and serene space, that once led me, in a reckless moment, to call the western public lands part of the geography of hope?[35]

Ecosystem management, moreover, asks our land management agencies to coordinate their management across administrative boundaries. It had become customary for natural resource agencies to view their boundaries as inviolate castle walls from which they defended the ramparts. Today these agencies are acknowledging that their boundaries are porous—that ecosystems do not observe our administrative lines upon the land. Ecosystem management requires that agencies and their neighbors, whether they be rural subdivisions or other land management agencies, work cooperatively. This new approach may help dim the artificial lines we have placed across the American landscape.

The signs are clear that we are meeting the present challenge. There will be a New West—one quite different from the old, and one that may very well last as long. It is being made today, in the region and on the land, by people who feel what Stegner meant when he wrote: "It would promise us a more serene and confident future if, at the start of our sixth century of residence in America, we began to listen to the land, and hear what it says, and know what it can and cannot do."[36]

Through his writings Wallace Stegner left us a road map to America, regardless of what region we call home. In his essay "Last Exit to America," he captured both the pathos of life today in this country as well as the hope. First the pathos:

> In Indiana, as you enter Interstate 80-90, the toll booth attendant will hand you a Travelaide packet containing strip maps, a directory of accommodations indexed by interchange, a list of scenic, historic, and cultural attractions on the road west, and the cheerful advice to "discover America." The information is reliable. The gas stations are where they are said to be, the accommodations are as advertised. But after a day and a half or so the traveler will realize that crossing the continent by Interstate he gets to know his country about as well as a cable message knows the sea bottom. It is not America that you pass through. You get a false impression even of the topography because the contours have been flattened, the grades leveled, gulches filled, hills cut away. It is clear from the Interstate that America is a coun-

> try of under three per cent grades. Regional cultures, local differences of human type, life style, dialect, diet? The right of way is wide enough to keep all that at a distance. The tourist dollar is lifted not by competing entrepreneurs but by franchise motels and restaurant chains which make every accommodation the replica of the last one—efficient, clean, stereotyped, and sterile.[37]

And now the hope:

> There was a man who did not have to run from his home canyon, whose house was contentment and sanctuary rather than restlessness and disgust, who consumed no transportation and pursued no pleasures beyond the pleasures of work and family and tribal association. There was a man who never had to learn how to give himself to his land of living.
>
> I saw him in the shade of a cliff, by a water-seep green with maidenhair and redbud, examining his simple, enviable life. I know he examined it: the tree growing from the protective hand, that hummingbird vivid in the tree, proved it. The picture he left behind him lured me toward resolutions I would probably not be able to keep. But I thanked him just the same.[38]

The solutions to the West's problems will not come from beyond the region; they must come from within. The history of the Mountain West has been a history of the slow and painful adaptation of societies and cultures to the western environment of rugged mountains and aridity. From the beginning, this process of adaptation has challenged westerners to create social forms appropriate to the situation. When these adaptations have worked, it was because they were cooperative and collectivist. The myth of rugged individualism was little more than just that and has probably caused more harm than good in the efforts of the West's inhabitants to live within the constraints imposed by the region. Stegner knew the western myths well enough to appreciate that they were inadequate: people could not live according to those myths without damaging the very land that, if cherished, could sustain societies that would be strong and vibrant.

The power of Stegner's writings was not due simply to the fact that he knew how to write. His writings were testimony to the strong love he felt for this region of breathtaking beauty and diversity. His books and essays have described a landscape whose very name, The West, causes people to pause. They have left us, too, with his expectations of how we might treat the land. Few people saw the West change so much, thought so long about these changes, and left so remarkable a written record of their thoughts as Wallace Stegner. Let him have the last word:

> I hope these essays do not say that western hopefulness is a cynical joke. For somehow, against probability, some sort of indigenous, recognizable culture has been growing on western ranches and in western towns and even in western cities. It is the product not of the boomers but of the stickers, not of those who pillage and run but of those who settle, and love the life they have made and the place they have made it in. There are more of those, too, than there used to be, and they know a great deal more, and are better able to resist and sometimes prevent the extractive frenzy that periodically attacks them.
>
> I believe that eventually, perhaps within a generation or two, they will work out some sort of compromise between what must be done to earn a living and what must be done to restore health to the earth, air, and water. I think they will learn to control corporate power and to dampen the excess that has always marked their region, and will arrive at a degree of stability and a reasonably sustainable economy based on resources that they will know how to cherish and renew. . . . I think they will do it. The feeling is like the feeling in a football game when the momentum changes, when helplessness begins to give way to confidence, and what looked like sure defeat opens up to the possibility of victory. It has already begun. I hope I am around to see it fully arrive.[39]

Although he is not here to witness it, Wallace Stegner, as much as anyone, showed us the direction toward a future where his legacy will someday be realized.

CHAPTER THIRTEEN

STEGNER AND CONTEMPORARY WESTERN POLITICS OF THE LAND

DOROTHY BRADLEY

THERE IS A DIRECT LINE from Aldo Leopold, who broadened our notions of ethics to include the land community, to Wallace Stegner, who adapted that land ethic to fit the West, to the sometimes thoughtful and frequently passionate attempts in Montana to incorporate these philosophies in real-life policy. Over the past quarter of a century, I have had much personal involvement with public life in Montana. During this period, I have observed and participated in the often wrenching process by which we seek to build an enduring American land ethic in our state.

When I was twenty-two I was unexpectedly called to Madison, Wisconsin, to the bedside of my mother, who was fading after a long illness. During my month's stay in Madison, my father revisited his Wisconsin roots with me and introduced me to Estella Leopold, Aldo Leopold's widow, who gave me a copy of *A Sand County Almanac*. The fact that my parents and I read it aloud in the hospital room, hanging onto every word, made it strikingly powerful. The fact that it became my spiritual platform when I ran for the Montana legislature seven months later made it sacred. And the fact that my father later married Nina (Estella and Aldo's first daughter) made it permanent.

While all of this is strictly personal, it was during this short period that I discovered my political wellspring. And Wallace Stegner's writings would keep it continuously recharged. Listening to Stegner on audiocassettes on my long road trips, mulling over his words, I would pull over, grab a pen, and write notes to myself on the political challenges of life in a land defined by aridity. Or on a wakeful night I would pull a Stegner volume from my bedside table and rethink what it means to belong to the land.

THESE MEMORIES are a prelude to a story involving Montana's environmental transformation. This particular story recalls events that unfolded nearly two and a half decades ago in response to the conditions created by a classic western economy reliant on fairly brutal treatment of the land.

Sometime around 1972, after I had served in my first legislative session, a small group of citizens, mostly ranchers, gathered in a cramped living room in Billings. We worried together about the overwhelming implications of coal development in eastern Montana. It was all very new. And it was alarming. Montana was beginning a hasty transition from the strip-mining of 3 million tons of coal per year to 30 million tons. Strip-mining removed the top few inches of soil, and a hundred feet or so of overburden, to shovel out the coal. What happens to the leftover landscape depends on the law.

We had in our hands a hefty federal document called the *North Central Power Study*. It raised the possibility that twenty-one new power plants would be built in our state—projects entailing multiple dams and aqueducts to provide the required cooling water and resulting in the stripping of thousands of acres of land. Someone had even coined a name for the area these projects would transform; they called it the "national sacrifice area."

We came to realize, at that homegrown meeting, that we needed a breather. The idea of a coal mining moratorium surfaced. We knew there were significant information gaps regarding the proposed development. The state had no inkling what we were in for, and no solid laws or policy for environmental protection. We needed time to learn and deliberate. We needed to put everything on hold until we, as public officials and (in a broader sense) as a state, had our act together. We needed to think through the water demands, the unknowns of reclaiming arid lands, the impact on groundwater (the lifeblood of eastern Montana), the siting of power plants and transmission lines, the air pollution, the social and economic effects, and the wishes of the landowners. Just the issue of private property rights alone was confusing. The surface of the land and the minerals beneath it had different owners—so whose rights should prevail? And Montana had frighteningly outdated eminent domain laws, left over from the days when the state provided an easy feast for the politically powerful copper kings of Butte and Anaconda. All in all, we were innocents. And so, in January 1973, we introduced a moratorium bill in the state legislature. I was its chief sponsor.

I remember that legislative year better than any other. Eastern Montana organized at the grass roots and sent one of the first presidents of their Northern Plains Resource Council to the legislature. Campaigning proudly as a "one-issue man," his sole legislative purpose was to pass a good strip mine reclamation act. His relatives had homesteaded a classic piece of Stegner Country—along Sarpy Creek in southeastern Montana (twelve inches of rain per year). He wore a moth-eaten sheepskin coat, he knew what he was talking about, and when he won the election we began to sense we were in for an interesting time.

The excitement intensified in the opening days of the legislature when the highly respected chairman of the appropriations committee and dean of the House, Francis Bardanouve, introduced his energy facility siting bill. Often bill sponsors must beg colleagues to join their crusade by signing on as cosponsors. Representative Bardanouve had no such problem. By popular demand he had to leave it on a table in the House chambers so that everyone could be a sponsor, flash their environmental credentials, and share the glory. (And this measure proposed a strict new regulatory scheme for the electrical generation process. It was no Mickey Mouse bill.)

The time had come for the moratorium to run the legislative gauntlet. The first hurdle was scheduling the public hearing. The clever Speaker of the House had sent the bill to the judiciary committee, bypassing the environmentally friendly natural resources committee. With little warning, the chairman, John Hall, scheduled the bill in the middle of the week. I was outraged. "It takes eastern Montana ranchers six hours to drive to Helena," I argued. "We need a weekend." As I made my emotional case, I wondered if my instincts were right—if masses of concerned citizens would indeed make the pilgrimage. The chairman reluctantly agreed.

The Saturday hearing was upon us and the Capitol was jammed. The committee, in fact the whole legislature, was stunned. In the early 1970s there was really no such thing as a citizens' lobby. Our lobbies were full of corporate interests and their expensive dark-suited lawyers. When the

Montana people moved into the Capitol that week of the moratorium, I remember one legislator protesting that we shouldn't allow all these arm-twisting, pushy citizens in—and they were lobbying without even having paid their ten-dollar fee! Another legislator told me he had disconnected his telephone because they wouldn't let him sleep at night.

The hearing was all you could wish for: real people—honest, sincere, and worried—making their case and asking for answers and help. A few days later the committee passed the moratorium bill to the House with "no recommendation," meaning that it came forward reflecting no sense of the sentiments of the committee—another extraordinary step. (Committee opinions are usually detailed and loud.)

On the night of the House debate, the large chamber was packed. People were overflowing from the galleries. No one had expected the bill to show its face outside the committee, yet here it was. We had no head count on our vote. We were all such novices, we hardly knew what a head count was. We didn't have much hope, either. We just had a very intense belief that this was the right thing to do.

The moratorium was one of those issues that was so precise and understandable that every legislator had to have his say. But then a waiting game began, for the former Speaker of the House, Jim Lucas, the most forceful orator I had ever heard, had not yet weighed in. He waited until late in the session to speak. When he rose, the whole House grew silent. Predictably and eloquently he spoke about how Montana must not fear to move forward; how eastern Montana was losing jobs and young people; about a new future for a hungry state. As he concluded, he lifted a huge stack of papers and booklets over his head, proclaiming that we already had all the studies we needed in hand, and dropped them in a pile on his desk. The time for studies was over, he told us. The time for courage and action was upon us. He sat down, and we sat dazed. We wondered numbly what the "studies" were.

And then Bob Watt, a seventy-five-year-old veteran legislator and retired teacher, rose slowly to his feet. The best it seems, was saved for last. Bob Watt, honest and thorough, was the very conscience of the legislature. Always accepting of innovative ideas, he and I, the oldest and youngest state representatives in Montana, had enjoyed cosponsoring legislation the previous session. In his remarks Bob cited the biblical call to subdue the earth—but not to beat it, trample it, ruin it. He berated the younger members of the House, noting the inappropriateness of his having to speak on behalf of future generations. The young people present, he suggested, had so much more to lose if Montana's heritage was needlessly sacrificed.

A light shuffle was heard in the chambers as, assuming the debate was over, we prepared to vote. Then a hush fell once more. John Hall, chairman of the judiciary committee, who had begrudgingly allowed us a Saturday hearing, was on his feet. John Hall—whom none of us believed cared about the Montana land or this piece of legislation—was recognized to speak. He stood a long time, silent, thinking. He stared at his desktop, just thinking. (I recall him often because he died unexpectedly just a few years later, and I never thanked him properly for the extraordinary thing he did next.) At last John spoke: of the magic of having the gift of a little time . . . the need for a brief pause in the scheme of things, to reflect . . . to help us create the right kind of future for our state, and all those who come after us. "Those who come after us. . . ." As he spoke, he gradually lifted the confusion from what we were trying to do. He brought respectability to the idea of a moratorium. It was a time to think.

When the vote went up for the count, the moratorium passed by a single vote.

In retrospect, I feel extraordinarily lucky to have been involved in the moratorium effort. Maybe the most important lesson is that political victories are seldom clear-cut. They come in different packages. We didn't win outright, of course. After a diligent night of quiet, powerful lobbying, the "old Montana" reasserted itself and the bill was killed. But we didn't lose, either. Rescued by a parliamentary procedure, the bill hung ominously over the legislature for the remainder of the session, like a thundercloud ready to rain, if the legislature didn't live up to its promises. Its presence shaped the remaining days of the legislative session. As a result, an abundance of protective bills passed. Politicians bragged for years afterward that Montana had the strongest environmental laws in the country (even years after it was no longer true).

THIS IS A STORY of transformation. Much came before, of course, laying the groundwork, and much has followed. But those were heady days, as Montana began to break away from its halcyon colonial traditions. Surely this is what settling in—belonging to a place—is all about. The West was casting aside the ghost-town, move-'em-out mentality; beginning to accept the realities of living dry; acquiring a taste for bunchgrass and sagebrush;

grasping the fact that we can't engineer dry into wet; and shifting from pillage to a perspective that reaches across generations, and not just annual financial statements. Maybe you could say that the West—or at least our corner of it—had finally begun to seek its angle of repose.

To this day, I carry around a little card, with words pecked out on an aging, portable, long-gone typewriter. It reads:

> The compelling fact is that the basic resources of water and soil, which can be mismanaged elsewhere without necessarily drastic consequences, cannot be mismanaged in the west without consequences that are immediate and catastrophic and that reach a long way.[1]

These are precise and stirring words written by—no surprise—Wallace Stegner. The responsibility he defined for us, and passed along to us, is a demanding task, if we seriously embrace its meaning.

Even those who would later advance the environmental agenda—reclamation, siting, water use, clean air, natural areas, and so on—frequently missed the point. When legislators argued into the night about reclamation land contours, native versus exotic vegetation, and bonding requirements, it was not so much the content that was significant as the collective *statement*: that the needs of future generations would no longer be left out of the equation. When we discussed siting transmission lines, the significance was not so much that for the first time in Montana history the "companies" had to compromise, but that we now recognized the external costs were immense, concealed, and permanent. And when Montana incorporated "natural areas" into its scheme for management of state-owned school trust lands, our policy for the first time was reflecting the conviction that long-term natural biotic diversity is as important to education as are short-term mining and timber dollars. Above all else, we were grasping the magnitude of our actions, talking about it, and accepting responsibility.

Political shifts do not happen unless they are preceded by carefully articulated ideas that have grown in enough hearts and minds to form a critical mass. The development of American conservation reflects, of course, a wonderful legacy involving a long roster of key historical figures. But I believe there are many involved in public life in the West who, like me, found Wallace Stegner to be the translator of the land ethic for their piece of the country—people who carry around little scraps of paper in their pockets with his words on them, reading them as both their compass and their inspiration. I can't tell you what a relief it was to discover, for example, that I didn't have to own my personal acre of Montana, because Montana owns me. This understanding has a name. It is known as "a sense of place."

As for our land community, and the task of ensuring its health and beauty for future generations, I have grave concerns that we are losing ground faster than the land ethic is advancing. Certain obstacles remain: Charles Wilkinson's "Lords of Yesterday," outdated laws, disorganized institutions. Even as our environmental crises are accelerating, it seems that our civic environment is crumbling and our collective patience for the complexity of issues and diversity of opinions is wearing thin. If there is a breakdown of our political process, how can we contemporize our tradition and update the law to reflect modern sentiments?

That is a discussion for another day. But since contemplations of Wallace Stegner focus us on the native land of hope, I would be remiss if I failed to acknowledge the watershed-oriented citizen organizations now cropping up in so many places, where those who love their tiny piece of the planet are going to great lengths to come together and negotiate to guide its future. Each generation is given its chance to build a better place, to leave its mark, "to define and illustrate the worthy life." Maybe we are truly grasping the possibilities of cooperation over rugged individualism in our striving to create a society more worthy of its remarkable landscape.

I would be remiss, too, if I failed to urge the great thinkers, teachers, and creative minds among us to carry on the torch. Each generation is indeed given its chance, but each generation needs its own manuals and tools, philosophers and poets. We should not take leave of Wallace Stegner without making a promise: not only to pass along his legacy, but to continue his quest—to clarify the task, translate the past, define the next meridian, and articulate the hope.

CODA

RECALLING STEGNER

In the concluding essay of this collection, Terry Tempest Williams recognizes the connections between Wallace Stegner and Aldo Leopold—and the links between their work in the past and the work that remains unfinished. The wild places of the continent, and of the world, are under constant pressure even as the meaning of the wilderness idea is debated and the need for restoration grows. Williams' essay concerns the power of words, and especially the act of writing, in pursuing conservation. She follows the connection between "conversation" and "conservation" in seeking "an empathetic response to the world we inhabit." In his own written "responses," Stegner composed elegant dialogue—dialogue not only between people but between people and place. Williams reminds us, as Stegner so often did, that love is defined by respectful words and by faithful silences.

CHAPTER FOURTEEN

WILDERNESS CONVERSATION

TERRY TEMPEST WILLIAMS

We gathered in Wisconsin to honor Wallace Stegner, to explore and celebrate his work, ideas, and life. The import and impressions of our gathering were inextricably linked with the opportunity to visit the sand county farm of Aldo Leopold. Sandhill cranes rose and fell over the marsh adjacent to Leopold's shack, as we stood on Leopold's land. Such a gift. In Leopold's day no more than twenty-five nesting pairs remained in Wisconsin. Now thousands of cranes are counted there every spring. Isn't this part of Wallace Stegner's "geography of hope"? Wherever the sacred is to be found, be it a person, a bird, a place, we recognize it as a homecoming. Family. Community.

During our time in Wisconsin I could not separate the legacy of Leopold from the legacy of Stegner. One gave us a land ethic, the other a wilderness credo. Imagine a dialogue between these two men, using corresponding passages from *A Sand County Almanac* and *The Sound of Mountain Water*.[1]

Wilderness is the raw material
out of which man has hammered
the artifact called civilization.

Wilderness—a genetic reserve,
a scientific yardstick by which

we may measure the world in its
man-made imbalance.

The rich diversity of the world's
cultures reflects a corresponding
diversity in the wilds that gave
them birth.

I want to speak for the wilderness
idea as something that has helped
form our character and that has
certainly shaped our history as a
people.

The most important characteristic
of an organism is that capacity for
internal self-renewal known as health.

We need wilderness preserved, as
much of it as is still left, and as
many kinds, because it was the
challenge against which our
character as a people was formed.

Ability to see the cultural value
of wilderness boils down, in the
last analysis, to a question of
intellectual humility.

We simply need that wild country
available to us—even if we never
do more than drive to its edge and
look in. For it can be a means of
reassuring ourselves of our sanity as
creatures, a part of the geography of
hope.

Two sandhill cranes dancing in the meadow, "trumpets" in the bog. Leopold and Stegner? A conversation about wilderness in Wisconsin, in the sand counties.

I did not know Aldo Leopold, but his words burn in me. His ethic lives in his children, in prairie seeds planted, in trees that grew, in the forest that now stands as a testimony to the work of restoration, a healing upon the land.

I did know Wally, and I loved him. Like so many, I still grieve his death.

I miss him, but he left us his words, his stories. This is our inheritance, a source of great joy and solace and wisdom.

Once when I was in trouble with the Mormon Church, Wally wrote me a letter. We had discussed the difficulty of having a free voice in Zion. His letter read, "Consider me your patriarch now. Whatever blessings you are in need of I will shower upon you. I am your Elder. Love, Wally."

He will forever be.

As his son Page has stated, "We are all his children." Yes. Family. Community. Braiding together, through our struggles and imagination, "a sense of place" and "a land ethic."

I believe we honor our elders by seeing them come to life through our actions. The shape of their mind, the cast of their soul, direct our own. Their spiritual influence still circulates through the world. Like the great cranes flying, their wings arch over us and we feel their shadows. We remember them. They remind us of what is possible.

Conduct . . . moral bearings . . . wit . . . savvy. . . . Wally always reminded us not to take ourselves so seriously, but to take the work very seriously. Their lives, Leopold and Stegner, create a compass of words and actions, pointing us in the direction of an empathetic response to the world we inhabit.

In 1960, Wallace Stegner wrote in his original "Wilderness Letter":

> Our western deserts are scarred somewhat by prospectors but are otherwise open, beautiful, waiting, close to whatever God you want to see in them. . . . [Robber's Roost, in Wayne County, Utah] is as lovely and terrible a wilderness as Christ and prophets went into; harshly and beautifully colored, broken and worn until its bones are exposed, its great sky without a smudge or taint from Technocracy, and in hidden corners and pockets under its cliffs the sudden poetry of springs. Save a piece of country like that intact, and it does not matter that only a few people a year will go into it. They can look two hundred miles clear into Colorado; and looking down over the cliffs and canyons of the San Rafael Swell and Robber's Roost they can also look as deeply into themselves as anywhere I know. And if they can't even get to the places on the Aquarius Plateau where the present roads will carry them, they can simply contemplate the idea, take pleasure in the fact that such a timeless and uncontrolled part of the Earth is still there.[2]

Stegner's pen issued words that elucidate and illuminate the civic responsibility of an artist. He was more than a social commentator, dramatizing his beliefs through fiction. He understood writing as "an act of consequence."

The conservation biologist E. O. Wilson speaks of "biophilia"—our need and affinity for life, the hunger we hold for affiliation with Other. He speaks also of "consilience," a unification of ideas at the intersection of biology, social science, ethics, and environmental policy.

Consilience. Inner and outer ecology. Home. Seeing the world whole, even holy. This is the work before us.

In the late 1980s, the Southern Utah Wilderness Alliance (SUWA), a small grassroots environmental organization, spearheaded the development of a document called "Wilderness at the Edge" in which thirty-five Utah conservation groups calling themselves the Utah Wilderness Coalition conducted their own wilderness inventory and made recommendations regarding a total of 5.7 million acres of Bureau of Land Management lands.

Wallace Stegner, a member of SUWA's board of directors, agreed to write the introduction for this citizen's handbook. His last paragraph reads:

> The Utah deserts and plateaus and canyons are not a country of big returns, but a country of spiritual healing, incomparable for contemplation, solitude, quiet, awe, peace of mind and body. We were born of wilderness, and we respond to it more than we sometimes realize. We depend upon it increasingly for relief from the termite life we have created. Factories, power plants, resorts we can make anywhere. Wilderness, once we have given it up, is beyond our reconstruction.[3]

Five years later, Utah's Republican representatives introduced in Congress the Utah Public Lands Management Act of 1995. Under the act, only 1.8 million out of 22 million acres of BLM lands were to be preserved—with language that encouraged construction of dams and communication lines and exploration for oil and gas. It undermined the Wilderness Act of 1964. The citizen's proposal was ignored.

How do we speak in this context? What would Stegner do? What would Leopold say?

I can tell you that citizens frustrated by the political process in Utah held their own hearings at the Indian Walk-in Center in Salt Lake City. Instead of three minutes, with the clock running and the delegation saying repeatedly "Sit down please" or "Please control yourselves" or "Please don't speak with passion" or "We will not have any philosophies discussed, only acreage," we established our own rules. People spoke from their hearts for as long as they wished. We were there until two in the morning.

Citizen representatives later testified at regional hearings brought to the state by the House Subcommittee on Natural Resources, chaired by Utah Congressman Jim Hansen. Maurice Hinchey (D–NY) sat on one side of him, Bruce Vento (D–MN) on the other. A personal anecdote serves to illustrate what we are up against. I was given the opportunity to testify at the hearing. I tried to speak as rationally as possible, representing the citizens' voices we had heard the night before. I tried to speak in a language that was credible and appropriate to the situation.

As I began speaking, I watched Congressman Hansen's eyes glaze over. He yawned, and I saw him take out his newspaper and start reading. In desperation, I broke from the text and said, "Congressman Hansen, I have lived here all my life. Is there anything I can say to you that might open your heart or mind to another alternative regarding these public lands?"

He leaned forward from the riser and said, "Excuse me, Mrs. Williams, there is something about your voice I cannot hear."

I do not believe he was referring to the microphone.

What do we do, each of us in our own place, in our own time?

In the midst of this debate, Mary Page Stegner phoned to say that she had been going through some of Wally's papers and had found a handwritten passage from one of his field notebooks in 1966 that might be helpful to us. It began with the simple question "Why?":

> "Why?" So people coming up dammed Glen Canyon in power boats can get out of the water and make it more easily from ramp to ramp. And anyone on the fringes of that lovely stone wilderness will hear their motors ten, fifteen, twenty miles away across the garbed stone, reverberating off the Kaiparowits Cliffs, ricocheting off the Waterpocket Fold.
>
> As there has been nothing to interrupt the silence in this desert, so is there nothing to interrupt the noise. As there is nothing to break the view of watercourses, of cliff lines and gulch and bare bald heads and domes, so there will be nothing to interrupt the view of cut and fill highway.
>
> This road as proposed by the Utah Highway Commission would be a tragedy, the dimmest of "wilderness breaking." Poverty program, public works. Yes—poverty of intelligence, poverty of imagination, poverty of vision, poverty of sensibility. And a greater poverty for Utah's future, once that last wilderness is split, shattered, and brought down to size.[4]

We have our elders, and they remind us of what is possible.

Stephen Trimble, another writer from Utah, and I met for lunch to discuss the situation before us. We remembered Stegner's 1960 letter to David Pesonen of the Outdoor Recreation Resources Review Commission, in which he addressed "the wilderness idea, which is a resource in itself." His "Wilderness Letter" gave us inspiration.

In early August 1995 we sent a letter to twenty-five writers who have Utah's redrock sand and alkali dust in their souls. The letter began simply, "Dear friends: We need your help." We asked them if they would consider writing a short essay about Utah's canyon country that we could incorporate into a chapbook that would be placed on the desk of every United States senator and representative.

Miraculously, three weeks later, we had nineteen essays in hand. We sought money from a local foundation, found a designer, and by the middle of September we had a small book entitled *Testimony*. We printed one thousand copies for distribution to all members of Congress, to the conservation community, and to the press.

With *Testimony* in hand, we flew to Washington, D.C., for a press conference at the Triangle in front of the Capitol on September 27, 1995. Several writers handed the book to Representatives Hinchey and Vento, who accepted *Testimony* on behalf of their colleagues and agreed to distribute the writings to each member of the House with a cover letter. Senator Russell Feingold (D–WI) subsequently sponsored *Testimony* in the Senate and sent the book to his colleagues with a similar letter.

After the press conference a journalist asked, "What will you do if the book makes no difference, if no one listens?"

"Writers never know the effect of their words," we responded. "We write as an act of faith."

Stegner will forever be our mentor. His words will always remind us of the artist's civic responsibility. Listen to the voices of Wallace Stegner's children.

Ann Weiler Walka:

> The rich blue mystery of the American continent, the dreamscape that harbors bears and cougars and mud-breathing fishes, informs and nurtures our collective soul. Settlers must have nodded ruefully when a Salt Lake paper reported that wilderness was good for nothing except to hold the world together. In fact, it does exactly that.[5]

Barry Lopez:

> We need a pause the like of which we've never had in Western civilization. We need to halt at watershed junctures like this one involving the disposition of Utah's public lands and ask not just what is fair, just, and reasonable, but what is enduringly wise.[6]

Rick Bass:

> You cannot convert the fragile stick nests of herons into timber or oil or gas. You cannot turn the whistle of wind across ancient orange sands into dollar bills or boats or security. You cannot cut a road into red rock across a sand creek and convert that loss into gain. This is and always has been a myth of mankind of all countries, both savage and civilized. There is a point reached in all cultures, a point of saturation, where each blade cut weakens a place and the miracle of regeneration does not one day occur.[7]

Ann Zwinger:

> . . . an owl feather and a grasshopper wing, and a chip of obsidian tied up with the song of a spadefoot toad, my own medicine bundle from my own ceremonies of passage. The desert grants each of us our own understandings, charges us with the preservation of its messages.[8]

William Kittredge:

> As we destroy that which is natural, we eat ourselves alive.[9]

"Something will have gone out of us as a people if we ever let the remaining wilderness be destroyed," echoes Stegner.

"Wilderness is the raw material out of which man has hammered the artifact called civilization," answers Leopold once again.

We are their children. They are our elders who continue to inspire. Their words give us courage to put our love into action. Writing as an act of faith. Writing as an act of consequence.

Family. Community. Even the sandhill cranes. Every writer I know in the American West feels the weight of Wallace Stegner's hand on their shoulder.

And every time I sit down to write I hear his voice. "Be bolder. Be braver. Be true to what you know in your bones, and love this bedrock, beautiful American West."

NOTES

Introduction

1. Jackson Benson, *Wallace Stegner: His Life and Work* (New York: Viking, 1996); *Wallace Stegner: A Writer's Life* (Los Angeles: Stephen Fisher Productions and KCET, 1996); Page Stegner and Mary Page Stegner, eds., *The Geography of Hope: A Tribute to Wallace Stegner* (San Francisco: Sierra Club, 1996); Charles E. Rankin, ed., *Wallace Stegner: Man and Writer* (Albuquerque: University of New Mexico Press, 1996); *Catching the Light: Remembering Wallace Stegner* (Covelo, Calf.: Yolla Bolly Press, 1996).
2. Wallace Stegner and Richard W. Etulain, *Conversations with Wallace Stegner on Western History and Literature* (Salt Lake City: University of Utah Press, 1983), pp. 32 and 35.
3. Stegner recalled that *Fire and Ice* (New York: Duell, Sloan, and Pearce, 1941) reflected his impression that "in any choice between left and right, between communists and fascists, I was going to have to go down the middle, because neither one of them seemed to me an absolutely imperative alternative. . . . *Fire and Ice* [was] the culmination of my feelings in Madison that this plus-or-minus, left-or-right business was an artificial and rather hysterical dichotomy. That feeling in me may have been partly innocence, coming as I did from a totally nonpolitical environment. But it may also have been regional, my protected sense of being deep in the continent" (Stegner and Etulain, *Conversations*, pp. 32–33).
4. Wallace Stegner, *Where the Bluebird Sings to the Lemonade Springs: Living and Writing in the West* (New York: Random House, 1992), pp. 199–206.
5. Wallace Stegner, *On the Teaching of Creative Writing*, Connery Lathem, ed. (Hanover, N.H.: University Press of New England, 1988), p. 38.
6. Wallace Stegner, *The Uneasy Chair: A Biography of Bernard De Voto* (Salt Lake City: Peregrine Smith Books, 1988), p. 274.
7. Ibid., p. 34.
8. Wallace Stegner, *One Way to Spell Man: Essays with a Western Bias* (Garden City, N.Y.: Doubleday, 1982), p. 108.

CHAPTER 1
Wallace Stegner and the Shaping of the Modern West

1. Wallace Stegner, *Wolf Willow: A History, a Story, and a Memory of the Last Plains Frontier* (New York: Viking, 1966), pp. 18–19.
2. Wallace Stegner, *The American West as Living Space* (Ann Arbor: University of Michigan Press, 1987), p. 71.
3. Wallace Stegner, *Beyond the Hundredth Meridian: John Wesley Powell and the Second Opening of the West* (Lincoln: University of Nebraska Press, 1982), p. 338.
4. Wallace Stegner, *The Big Rock Candy Mountain* (Lincoln: University of Nebraska Press, 1983), pp. 463–464.
5. Walter Prescott Webb, *The Great Plains* (Boston: Ginn, 1931), p. 192.
6. Professors Bock and Armstrong made these observations during presentations at the annual conference of the Center of the American West in Boulder, Colorado, on March 16, 1996.
7. Stegner, *Wolf Willow*, p. 59.
8. Stegner, *The American West as Living Space*, p. 80.
9. Wallace Stegner, *The Sound of Mountain Water: The Changing American West* (New York: Dutton, 1980), p. 102.
10. Ibid., p. 18.
11. Page Stegner, "A Brief Reminiscence: Father, Teacher, Collaborator," *Montana: The Magazine of Western History* 43 (Autumn 1993): 55. Reprinted in Charles E. Rankin, ed., *Wallace Stegner: Man and Writer* (Albuquerque: University of New Mexico Press, 1996), pp. 27–33; Page Stegner and Mary Stegner, eds., *The Geography of Hope: A Tribute to Wallace Stegner* (San Francisco: Sierra Club, 1996), pp. 103–110.
12. Stegner, *Big Rock Candy Mountain*, p. 332.
13. Stegner, *Beyond the Hundredth Meridian*, p. 134.
14. Stegner, *Wolf Willow*, p. 85.
15. Stegner, *Sound of Mountain Water*, pp. 146, 148, and 153.
16. Stegner, ibid., p. 38.
17. Stegner, *Wolf Willow*, p. 306.

PART I
Stegner As Writer

1. See especially Patricia Nelson Limerick's essay "Precedents to Wisdom" in Charles E. Rankin, ed., *Wallace Stegner: Man and Writer* (Albuquerque: University of New Mexico Press, 1996), pp. 105–117.

CHAPTER 2
Writing As the Expression of Belief

1. Nancy Packer, untitled speech delivered at Wallace Stegner's eightieth birthday celebration, Buck House, Stanford University, April 30, 1989.
2. Wallace Stegner, *Where the Bluebird Sings to the Lemonade Springs: Living and Writing in the West* (New York: Random House, 1992), p. 222.
3. James D. Houston, interview with the author, January 10, 1990.
4. Robert Stone, *Outerbridge Reach* (New York: Ticknor and Fields, 1992), p. 338.
5. Wallace Stegner, "Variations on a Theme by Conrad," *Yale Review* 39 (1950):521.
6. Wallace Stegner, "Robert Frost: A Lover's Quarrel with the World," *Stanford Today* 133 (1961).
7. David Dillon, "Time's Prisoners: An Interview with Wallace Stegner," in Anthony Arthur, ed., *Critical Essays on Wallace Stegner* (Boston: G. K. Hall, 1982), p. 57.
8. Leslie Hanscom, "I Write the Kind of Novel I Can Write" (interview with Wallace Stegner), *Newsday*, March 11, 1979, p. 18.
9. Kay Mills, "A Look at the Real West" (interview with Wallace Stegner), *San Francisco Examiner*, August 20, 1977.
10. Hanscom, "I Write the Kind of Novel I Can Write," p. 18.
11. James R. Hepworth, "The Art of Writing: An Interview with Wallace Stegner," *Bloomsbury Review* 10(2) (March/April 1990):9.
12. Wallace Stegner and Mary Stegner, "Introduction," in Wallace Stegner and Mary Stegner, eds., *Great American Short Stories* (New York: Dell, 1957), p. 15.
13. Henry James, *The Wings of the Dove* (New York: Dell, 1965), p. 512.
14. Wallace Stegner, *Angle of Repose* (Garden City, N.Y.: Doubleday, 1971), p. 569.
15. Wallace Stegner and Richard W. Etulain, *Conversations with Wallace Stegner on Western History and Literature* (Salt Lake City: University of Utah Press, 1983), p. 64.
16. Stegner and Stegner, *Great American Short Stories*, p. 16.
17. Stegner and Etulain, *Conversations*, p. 64.
18. Kay Bonnetti, *Interview with Wallace Stegner*, Audio Prose Library, February 1987.
19. Stegner and Etulain, *Conversations*, p. 78.
20. Stegner, *Where the Bluebird Sings*, p. 219.
21. Ibid., p. 217.
22. Patricia Anderson, Walter Rideout, and Gretchen Schoff, interview with Wallace Stegner, University of Wisconsin, May 1986.
23. Wallace Stegner, "Sensibility and Intelligence," *Saturday Review*, December 13, 1958, p. 24.

24. Wallace Stegner, "Bugle Song," in *Collected Stories of Wallace Stegner* (New York: Penguin, 1991), p. 13.

CHAPTER 3
Wallace Stegner's Hunger for Wholeness

This essay first appeared in the August 1997 issue of the *High Plains Literary Review.*

1. Wallace Stegner, *The Sound of Mountain Water: The Changing American West* (New York: Dutton, 1980), pp. 41–42.
2. Wallace Stegner, *Wolf Willow: A History, a Story, and a Memory of the Last Plains Frontier* (New York: Viking, 1962), pp. 1–2; subsequent pages are indicated in the text.
3. Stegner, *Sound of Mountain Water*, p. 210.
4. Wallace Stegner, *Where the Bluebird Sings to the Lemonade Springs: Living and Writing in the West* (New York: Random House, 1992), p. 218.
5. Ibid., p. 32.
6. Kay Bonnetti, *Interview with Wallace Stegner*, Audio Prose Library, February 1987.
7. Wallace Stegner, *One Way to Spell Man: Essays with a Western Bias* (Garden City, N.Y.: Doubleday, 1982), p. 156.
8. Stegner, *Where the Bluebird Sings*, p. 10.
9. Bonnetti, *Interview with Wallace Stegner.*
10. Stegner, *One Way to Spell Man*, p. 141.
11. Wallace Stegner and Richard W. Etulain, *Conversations with Wallace Stegner on Western History and Literature* (Salt Lake City: University of Utah Press, 1983), p. 197.
12. Stegner, *One Way to Spell Man*, p. 67.

CHAPTER 4
Ruminations on Wallace Stegner's Protective Impulse and the Art of Storytelling

1. Wallace Stegner, *Crossing to Safety* (New York: Random House, 1987), p. 139.
2. Ibid., p. 51.
3. Wallace Stegner, *Where the Bluebird Sings to the Lemonade Springs: Living and Writing in the West* (New York: Random House, 1992), p. 130.
4. Wallace Stegner, *The Big Rock Candy Mountain* (New York: Penguin, 1991), pp. 124–126; subsequent pages are indicated in the text.
5. Stegner, *Where the Bluebird Sings*, p. 227.

6. Wallace Stegner and Richard W. Etulain, *Conversations with Wallace Stegner on Western History and Literature* (Salt Lake City: University of Utah Press, 1983), p. 43.
7. See John Daniel, "Wallace Stegner's Hunger for Wholeness," Chapter 3 in this volume.
8. Stegner, *Where the Bluebird Sings*, p. 24.
9. Ibid., p. 24.
10. Ibid., p. 25.
11. Stegner and Etulain, *Conversations*, p. 43.
12. Wallace Stegner, *Recapitulation* (Lincoln: University of Nebraska Press, 1986), p. 137.
13. Stegner and Etulain, *Conversations*, p. 47.
14. Levette Jay Davidson, "Letters from Authors," *Colorado Magazine* 19(3) (July 1942), 123.
15. Wallace Stegner, *Angle of Repose* (Garden City, N.Y.: Doubleday, 1971), p. 239.
16. Ibid., p. 536.
17. Stegner's use of the Foote materials—quoting nearly verbatim passages from her letters and reminiscences and constructing his novel around metaphors and images ("angle of repose" is one) she used in her own writing—has been a source of some controversy among readers of both Stegner and Foote. For more information see Mary Ellen Williams Walsh, "Angle of Repose and the Writings of Mary Hallock Foote: A Source Study," pp. 184–209 in Anthony Arthur, ed., *Critical Essays on Wallace Stegner* (Boston: G. K. Hall, 1982). As the present discussion of Foote's work suggests, she deserves much wider attention as a fine western writer and illustrator whose work in many ways anticipates themes of New Western History; she wrote, for instance, about water rights, the effect of eastern capital on the development of the West, and mining and engineering. Stegner was one of the first writers to give her that attention by teaching her work in his classes at Stanford and reprinting her stories. Critical response to *Angle of Repose* has also drawn more attention to her work. Exploring the ways in which Stegner used his historical sources, and departed from them, can help us understand the novel more fully.
18. See Melody Graulich, "The Guides to Conduct That a Tradition Offers: Wallace Stegner's *Angle of Repose*," *South Dakota Review* 23 (Winter 1985):87–106; and "Book Learning: *Angle of Repose* as Literary History," in Charles E. Rankin, ed., *Wallace Stegner: Man and Writer* (Albuquerque: University of New Mexico Press, 1996), pp. 231–253.
19. Stegner, *Angle of Repose*, p. 524.
20. Ibid., p. 540.
21. Ibid., p. 475.
22. Ibid., p. 534.
23. Stegner, *Where the Bluebird Sings*, p. 25.

24. Mary Hallock Foote, *The Last Assembly Ball and The Fate of the Voice* (Boston: Houghton Mifflin, 1889), pp. 39–40.
25. Stegner, *Angle of Repose*, p. 540.
26. Wallace Stegner, *The Sound of Mountain Water: The Changing American West* (New York: Dutton, 1980), p. 181.
27. Stegner, *Crossing to Safety*, p. 196.
28. Stegner, *Where the Bluebird Sings*, pp. 226–227.
29. Stegner, *Recapitulation*, p. 25.
30. Stegner, *Sound of Mountain Water*, p. 147.
31. Stegner, *Where the Bluebird Sings*, p. 151.
32. Terry Tempest Williams, *Refuge* (New York: Vintage, 1992), p. 119.
33. Stegner, *Where the Bluebird Sings*, p. 218.
34. Wallace Stegner, *The Spectator Bird* (Lincoln: University of Nebraska Press, 1979), p. 213.

CHAPTER 5
Wallace Stegner's Practice of the Wild

1. James R. Hepworth, "Wallace Stegner's *Angle of Repose*: One Reader's Response" (Ph.D. dissertation, University of Arizona, 1989). The best-known collection of interviews is Wallace Stegner and Richard Etulain, *Conversations with Wallace Stegner on Western History and Literature* (Salt Lake City: University of Utah Press, 1983); reprinted as *Stegner: Conversations on History and Literature* (Reno: University of Nevada Press, 1996). Readers interested in my interviews with Stegner should consult the appendixes to my dissertation. Excerpts have been published in *The Bloomsbury Review* (March/April 1990), *The Paris Review* (Summer 1990), and *Stealing Glances: Three Interviews with Wallace Stegner* (Albuquerque: University of New Mexico Press, forthcoming).
2. Gary Snyder, *The Practice of the Wild* (San Francisco: North Point, 1990), p. 16. My essay draws heavily on Snyder's definition of wilderness in *The Practice of the Wild*, especially the essay entitled "The Etiquette of Freedom" (pp. 3–24). So far, the only discussion of the relationship between Wallace Stegner and Gary Snyder appears in Jackson Benson's biography *Wallace Stegner: His Life and Work* (New York: Viking, 1996) in the chapter entitled "Trouble in the Sixties." Because Stegner's readers still tend to view him as a provincial thinker and confuse him with his narrators—especially "establishment" reactionaries like Joe Allston and Lyman Ward—I'm hoping to help mend a few fences in this essay. In my opinion, Stegner's impatience with the "counterculture," I think, has been overrated. Insofar as it did exist, it was rooted in his own steadfast belief in moderation and self-control, but especially in his cyclical view of history. As early as 1958 Stegner had written a defense of

art and artistic freedom, mostly as a critical assault on the nation's all-consuming love affair with science, "progress," and the empirical method. That essay ("One Way to Spell Man") speaks forcefully against the prevailing view that the "arts and literature are charming frills," but [mere] "hangovers from the days of witchcraft and wonder." Science, in this view, will sooner or later render the arts unnecessary. Stegner, by contrast, believed that the arts and sciences are complementary and even supplementary, that the arts are the cultural carriers of truths that science is hopelessly inadequate to explain, and that artistic rebellions are necessary episodes that imitate and repeat each other throughout the evolution of civilizations.

Stegner, I believe, viewed the cultural revolutions of the 1950s, 1960s, and 1970s as necessary reactions to the cultural malaise of the nation that he had diagnosed early in books like *One Nation* and *Beyond the Hundredth Meridian.* Certainly he believed in tradition—in the great community of human thought. Consequently he feared "immoderate zeal" and held "passionate faith" suspect. On the other hand, Stegner never lost sight of the need for artists to rebel against their traditions. He was himself a rebel. But his rebellion was against the anti-educational, anti-intellectual position of his father and, consequently, toward history and tradition. It was a rebellion, too, against literary fashion and experimentalism for the sake of experimentalism. As an artist, however, Stegner was all his life an "outsider." In the introduction to his biography, Benson is eloquent on this point, which is a subtle and paradoxical notion to convey. But Stegner was nevertheless quite clear about the artist's need to rebel. Consider, for example, the following paragraph:

> But if art is in good part tradition, its truths, being linked with response, are the kind that will vary in kind or intensity with different responders. More than that, the most extreme rebellions by artists against their traditions are both possible and desirable. Picasso's violences against perspective, like Schoenberg's against key signatures and Faulkner's against syntax, are as radical as if in medicine some research doctor had declared against the germ theory of disease, and yet they are vital, healthy, and necessary. They assert the fullest freedom to question experience in any terms. [*One Way to Spell Man: Essays with a Western Bias* (Garden City, N.Y.: Doubleday, 1982), p. 14.]

Although Stegner was impatient with the drug culture and the counterculture's self-righteous and extreme positions, he nevertheless came to view Gary Snyder as a consummate artistic rebel. Moreover, the point is that Stegner and Snyder bridge a gulf between generations. Although differences certainly count, their views of wilderness, art, culture, race, gender, love, and marriage are far more compatible than antagonistic. Although Benson doesn't say so, it appears almost certain that Stegner read an early copy of Gary Snyder's essay "Four Changes" (1969). From this Snyder piece Stegner cut snippets and integrated them into "The Mesa" section of *Angle of Repose.* It appears almost equally certain that Snyder was not amused. The two of them had already exchanged letters in 1968 on the topic of "real values" and Snyder's Stanford speech. Regardless, when during a break in our first interview in 1977 I commented to Stegner that I found his use of Snyder's words in the novel both ironic and comic, he said, "I'm not sure Gary would agree with you." Then he smiled. During the interview I had raised a question about "the incorporation of the traditional thought of the Far East in current American thought," and Stegner had responded as follows:

> What's happening is that the culture is changing, and when the culture changes, this will inevitably be reflected in the fiction. Maybe the fiction will have some part in changing the culture. I don't know that our culture can't contain those things. It's absolutely open, and there are plenty of people spending their time studying Zen and other traditions. Gary Snyder is a good case in point. I suppose Gary Snyder makes Zen and Old Coyote and other kinds of people come together in a reasonable way. On the other hand, it can go to the point of absurdity, too, and you can get the Moonies incorporating their thing into the culture. I think that just has to be worn out and thrown away. [Hepworth, "One Reader's Response," p. 498]

Snyder himself has been open in his admiration for Stegner's conservation work. When *The American West as Living Space* (Ann Arbor: University of Michigan Press, 1987) first appeared, Snyder cited it as one of the best books of the year in the op-ed section of *The Bloomsbury Review*. Once he accepted his teaching position at the University of California at Davis, Snyder invited Stegner to the campus, though Stegner had to decline. In fact, Snyder and his publisher, Jack Shoemaker, had planned to visit Stegner at his home the same spring that Stegner died. In his new book, *A Place in Space*, Snyder dedicates an essay to Stegner. Its title, "The Rediscovery of Turtle Island," echoes the title of one of Stegner's own essays, "The Rediscovery of America: 1946." In that essay Stegner had let himself be seduced by the notion of progress. He writes admiringly of Hoover Dam and Lake Mead. Looking back over his words while compiling his first collection of essays in 1968 and then again in 1980 when *The Sound of Mountain Water* was reissued, Stegner cringed. But he let the essay stand as a scar to mark the evolution in his thinking about hydroelectric power—perhaps to remind both himself and his readers that experience is always developing, and if he could alter his thinking, so too could the rest of us.

3. Wallace Stegner, *The Sound of Mountain Water: The Changing American West* (New York: Dutton, 1980), p. 153.
4. Wallace Stegner, *Angle of Repose* (Garden City, N.Y.: Doubleday, 1971), p. 17.
5. Wallace Stegner, *Crossing to Safety* (New York: Random House, 1987), p. 3.
6. Wallace Stegner, *All the Little Live Things* (New York: Viking, 1967), p. 21.
7. Stegner, *Sound of Mountain Water*, p. 148.
8. Wallace Stegner, *Where the Bluebird Sings to the Lemonade Springs: Living and Writing in the West* (New York: Random House, 1992), pp. 210–211.

9. Ibid., p. 210.
10. Mary Ellen Williams Walsh raises the issue of plagiarism in connection with Stegner's *Angle of Repose* in *Critical Essays on Wallace Stegner* (1982), edited by Anthony Arthur, and Jackson Benson discusses the issue (and Walsh's treatment of it) at length in his biography *Wallace Stegner: His Life and Work.* In the title essay from her book, *Why I Can't Read Wallace Stegner* (Madison: University of Wisconsin Press, 1996), Elizabeth Cook-Lynn accuses Stegner of immorality in his treatment of Native Americans in *Wolf Willow*—an idea at odds with Elliot West's statement in Chapter 6 of this volume that *Wolf Willow* is the only case of Stegner's "bringing Indians into his narrative as flesh-and-blood, changing humans with long pasts, evolving identities, and, at least by implication, some kind of future."

Essentially, Walsh and Cook-Lynn use Stegner as their whipping boy to call attention to their own political agendas. During our recorded and unrecorded conversations for my dissertation and the "Writers at Work" interview for *The Paris Review*, I took up the topic of Native American history and literature with Stegner several times. He pointed me first to his book *One Nation*, which, he said, had been largely ignored since going out of print "and maybe deserves to be." Stegner's self-deprecatory words notwithstanding, as Patricia Nelson Limerick points out in her article "Precedents to Wisdom" (in Page Stegner and Mary Stegner, eds., *The Geography of Hope: A Tribute to Wallace Stegner* (San Francisco: Sierra Club, 1996), pp. 21–28; reprinted in revised form in Charles E. Rankin, ed., *Wallace Stegner: Man and Writer* (Albuquerque: University of New Mexico Press, 1996), pp. 105–117), Stegner knew "that the workings of human memory and loyalty are as funny as they are powerful" (p. 21). He "was just as dramatically ahead of his time in the matter of race relations as he was in environmental affairs" (p. 22)—and, one might also add, in matters of historiography as well. Indeed, Stegner was by definition a New Western Historian before any of the self-styled New Western Historians were born. (See the essays by West, Nugent, Flores, and Vale in this volume.)

Because I had read *One Nation* before I interviewed Stegner, I was able to press him on several of these matters, especially on the subject of Native Americans. I asked, for instance, why so few Indians figure in his histories and why none that I could recall ever showed up in his fiction. On the subject of fiction he was eloquent, and his position was essentially the one that Sherman Alexie often argues. Stegner said that he believed firmly in the young Indian writers and their ability to tell their own stories—to create a "usable past" in their histories and fictions, a past he thought had been largely misunderstood, overly romanticized, and mythologized practically out of existence. He also referred frequently to established Indian writers like N. Scott Momaday. Stegner said he had determined on his own and "pretty early" that Anglo-American writers of every description—including novelists, anthropologists, and historians—had co-opted "Indian Territory," and some of them had done so, he frankly stated, "carelessly and irresponsibly." He cursed the writers of pulp fiction and lauded Vine Deloria as an essayist and historian.

He pointed out, too, that few if any Afro-Americans enter into his fictional worlds, and even fewer Asian Americans and Hispanics. He hoped that wouldn't keep Afro-American, Asian American, and Hispanic-American readers from reading his books any more than their choices of subject matter and character and events kept him from reading theirs. Stegner also observed that "cultural matters are always much more complex than they appear to outsiders"—which is the reason, he said, that the game of literature is best played between readers and writers who are the products of similar cultures. He was, he said, no expert in the cultural matters of any of the Native American tribes. He and his novels were, he admitted, outside the American vanguard, "far from the cutting edge." As a fiction writer, he confessed to being firmly planted in realism and moralism. And he was "equally unfashionable" as a historian, he said, because he was committed to narrative history as "a branch of literature, not history as social science." "In a phrase," he said, "I am hopelessly culture-bound."

11. Stegner, *One Way to Spell Man*, p. 108.
12. Stegner, *Where the Bluebird Sings*, p. 200.
13. Ibid., p. 200.
14. Snyder, *The Practice of the Wild*, p. 10.
15. Stegner, *One Way to Spell Man*, p. 4.
16. Benson, *Wallace Stegner*, p. 303.
17. Ibid., p. 309.
18. Ibid., p. 314.
19. Wallace Stegner, *The Spectator Bird* (New York: Doubleday, 1976), p. 138.
20. Ibid., p. 146.
21. Stegner, *Sound of Mountain Water*, p. 147.
22. Stegner, *The Spectator Bird*, p. 105.
23. Snyder, *The Practice of the Wild*, p. 22.
24. Ibid., p. 23.
25. Ibid., p. 24.
26. Ibid., p. 24.
27. Stegner, *The Spectator Bird*, p. 52.
28. Ibid., p. 56.
29. Ibid., pp. 56–57.
30. Snyder, *The Practice of the Wild*, p. 28.
31. As Basso notes in *Wisdom Sits in Places* (Albuquerque: University of New Mexico Press, 1996), "The study of American Indian place names has fallen on hard times," although it was once "a viable component of anthropology" practiced by Franz Boas, Edward Spicer, and other preeminent anthropologists (p. 43). In fact, Basso's is the first major study in twenty-five years. In reviewing the book, N. Scott Momaday comments, "Place may be the first of

all concepts; it may the oldest of all words." In his text Basso quotes a passage from Vine Deloria Jr.'s *God Is Red* (1975): "American Indians hold their lands—places—as having the highest possible meaning, and all their statements are made with this reference point in mind." No one who thoughtfully reads Stegner's work, I think, could help but conclude that he shared this Native American perspective and in fact consciously attempts to apply it in his fiction. Yet no one, with the possible exceptions of Elliot West and Melody Graulich (in *South Dakota Review,* 1985), has paid much attention to Stegner's own process of naming and the methods he uses for building his own "place-worlds" (Basso's term). Basso cites Stegner as one of several authorities to whom he turned for help in negotiating what he considers "lightly chartered territory." His book is indispensable for anyone genuinely interested in Stegner's own sense of place or the relationship between land, people, and values. Readers interested in Stegner's naming process may also wish to consult my own beginning of the discussion in Hepworth, "One Reader's Response."

32. Snyder, *The Practice of the Wild*, p. 18.
33. Stegner, *The Spectator Bird*, p. 57.
34. Ibid., p. 58.
35. Ibid., p. 9.
36. Ibid., p. 152.
37. Aldo Leopold, *A Sand County Almanac and Sketches Here and There* (New York: Oxford University Press, 1949), p. 133.
38. Stephen Mitchell, *Tao Te Ching* (New York: HarperCollins, 1988), p. 29.
39. Stegner, *Where the Bluebird Sings*, p. 132.
40. Hepworth, "One Reader's Response," p. 500.
41. Wendell Berry, *Sex, Economy, Freedom & Community: Eight Essays* (New York: Pantheon, 1992), p. 101.
42. Stegner, *Where the Bluebird Sings*, p. 199.
43. Stegner, *One Way to Spell Man*, p. 22.
44. Stegner, *Sound of Mountain Water*, p. 146–147.
45. Stegner, *The Spectator Bird*, p. 213.
46. Snyder, *The Practice of the Wild*, p. 19.
47. John Gardner, *The Art of Fiction: Notes on Craft for Young Writers* (New York: Alfred Knopf, 1984), p. 18.
48. Stegner, *One Way to Spell Man*, p. 11.
49. The best book on the "nonfiction novel" I've found is John Hollowell's *Fact & Fiction: The New Journalism and the Nonfiction Novel* (Chapel Hill: University of North Carolina Press, 1977). Hollowell's book was the first to treat the subject in depth, but because it is out of print it has generally escaped notice. The relationship between Capote (who coined the phrase "nonfiction

novel") and Stegner receives a glancing reference in Benson, *Wallace Stegner*, p. 81.

50. Stegner, *One Way to Spell Man*, 76.
51. Tristine Rainer, *The New Diary: How to Use a Journal for Self-Guidance and Expanded Creativity* (Los Angeles: Tarcher, 1978), p. 28.
52. Ibid., p. 11.
53. Ibid., p. 18.
54. Ibid., p. 26.
55. Native American history (history told or recorded by Native Americans) is nearly uncharted territory for most Americans. In *Red Earth, White Lies: Native Americans and the Myth of Scientific Fact* (New York: Scribner's, 1995), Vine Deloria Jr. estimates that "only 10 percent of the information [concerning the Pre-Columbian history of the hemisphere] that Indians possess is presently in print and available for discussion" (p. 11). His book is perhaps the first to challenge the "obvious pattern of nonsense" that appears in much scientific writing—writing that depends upon very little other than the "authority" of the writer and America's almost unswerving faith in the myth of science. Deloria contends that recent attempts by "scientists" to hold Native Americans responsible for the "slaughter" of Pleistocene megafauna are empirically baseless.
56. For a discussion of Stegner and the oral tradition, see Hepworth, "One Reader's Response." Peter Farb's *Word Play: What Happens When People Talk* (New York: Knopf, 1974) is still a great introduction to the oral tradition for general readers. Lawrence J. Evers' "Native Oral Traditions," in *A Literary History of the American West* (Fort Worth: Texas Christian University Press, 1987), is a superb general guide to American Indian oral traditions. N. Scott Momaday's "Man Made of Words," in *Indian Voices: The First Convocation of American Indian Scholars* (San Francisco: Indian Historian Press, 1970), remains unsurpassed as an original and imaginative treatment of the Native American oral tradition; see also Momaday's "The Native Voice," in Emory Elliott, ed., *The Columbia Literary History of the United States* (New York: Columbia University Press, 1988), pp. 5–15. Albert B. Lord's *The Singer of Tales* (Cambridge, Mass.: Harvard University Press, 1960) is the definitive work on the oral epic in the literature of western civilization. Keith H. Basso's groundbreaking *Portraits of "The Whiteman": Linguistic Play and Cultural Symbols Among the Western Apache* (New York: Cambridge University Press, 1979) is a masterpiece of linguistic ethnography. It focuses on the way Native Americans—in this case the Cibecue Apaches—use the oral tradition to make sense of the behavior of Anglo-Americans. Dell Hymes, Dennis Tedlock, and Barbara Tedlock are among the few readable contemporary authors doing the necessary work to bring Native American oral literature and history to the attention of general readers.
57. William Stafford, *Writing the Australian Crawl: Views on the Writer's Vocation* (Ann Arbor: University of Michigan Press, 1978), p. 17.

58. The old saying "Intention endangers creation" is one the moderns took to heart. And in many ways Stegner, like his friend Ansel Adams, is a third-generation modern, who insists on taking an interdisciplinary approach to artistic endeavors and making use of "primitive" techniques and methods as well as new ones. The "found" objects in Stegner's fiction underscore his commitment to point of view and perspective as the literary artist's primary considerations. For an extended discussion of this topic and a reference to Stegner's "found objects" see Lynn Stegner's essay "A Point of View" in *The Geography of Hope*. The practice of incorporating a "found" manuscript into a novel, which Stegner frequently employs, is almost as old as the American novel itself and is perhaps best exemplified by Hawthorne in *The Scarlet Letter*, although the practice actually begins with the origins of the novel in England and writers like Richardson. Stegner uses the device as a metafictional technique in both *Angle of Repose* and *The Spectator Bird*.
59. William Blake, in Mary Lynn Johnson, ed., *Blake's Poetry and Designs* (New York: Norton and Company, 1979), p. 345.
60. Stegner, *One Way to Spell Man*, p. 8.
61. Walt Whitman, "Facing West from California's Shores," in Nina Baym et al., eds., *The Norton Anthology of American Literature*, 4th ed. (New York: Norton, 1994), vol. 1, p. 2025.
62. Stegner, *Sound of Mountain Water*, p. 183.
63. Ibid., p. 179.
64. Snyder, *The Practice of the Wild*, p. 15.
65. Stegner, *Sound of Mountain Water*, p. 146.
66. Ibid., p. 153.
67. Ibid., p. 152.
68. Hepworth, "One Reader's Response," p. 596.

Part II
Stegner As Historian

1. Wallace Stegner, *Wolf Willow: A History, a Story, and a Memory of the Last Plains Frontier* (Lincoln: University of Nebraska Press, 1966), p. 28.
2. See Wes Jackson, *Becoming Native to This Place* (Lexington: University Press of Kentucky, 1994).
3. Wallace Stegner and Richard W. Etulain, *Conversation with Wallace Stegner on Western History and Literature* (Salt Lake City: University of Utah Press, 1983), p. 166.
4. Gary Topping, "Wallace Stegner the Historian" in Charles Rankin, ed., *Wallace Stegner: Man and Writer* (Albuquerque: University of New Mexico Press, 1996), pp. 146–147. Rob Williams' essay "'Huts of Time': Wallace Stegner's Historical Legacy," in the same volume, discusses the problematic nature of Stegner's historical writing as well.

CHAPTER 6
Wallace Stegner's West, Wilderness, and History

1. Wallace Stegner, *Angle of Repose* (Garden City, N.Y.: Doubleday, 1971), p. 17.
2. Wallace Stegner, *The Sound of Mountain Water: The Changing American West* (Lincoln: University of Nebraska Press, 1985), p. 199.
3. George Perkins Marsh, *Man and Nature: Or, Physical Geography as Modified by Human Action*, David Lowenthal, ed. (Cambridge, Mass.: Harvard University Press, 1965), p. 36.
4. Karl Hess, *Rocky Times in Rocky Mountain National Park: An Unnatural History* (Niwot: University Press of Colorado, 1993), pp. 43–45.
5. John Muir, *My First Summer in the Sierra* (San Francisco: Sierra Club, 1988), 110.
6. Stegner, *Sound of Mountain Water*, p. 178.
7. Forrest G. Robinson and Margaret G. Robinson, "An Interview with Wallace Stegner," *American West* 15(1) (January/February 1978):36.
8. Wallace Stegner, *Where the Bluebird Sings to the Lemonade Springs: Living and Writing in the West* (New York: Random House, 1992), p. 201.
9. Stegner, *Sound of Mountain Water*, p. 91.
10. Stegner, *Where the Bluebird Sings*, p. 206.
11. Ibid., p. 5.
12. Stegner, *Sound of Mountain Water*, p. 199.
13. Ibid., p. 201.

CHAPTER 7
Wallace Stegner, John Wesley Powell, and the Shrinking West

1. Wallace Stegner, *Where the Bluebird Sings to the Lemonade Springs: Living and Writing in the West* (New York: Random House, 1992), p. 3; subsequent pages are indicated in the text.
2. Wallace Stegner and Richard W. Etulain, *Conversations with Wallace Stegner on Western History and Literature* (Salt Lake City: University of Utah Press, 1983), p. 148.
3. Ibid., p. 164; Etulain's words.
4. Ibid., pp. 165–166.
5. Ibid., p. 166.
6. Francis Parkman, *La Salle and the Discovery of the Great West* (New York: Library of America, 1983), vol. 1, chaps. 5–6.
7. Stegner and Etulain, *Conversations*, pp. 182–83.
8. Ibid., p. 181.
9. Ibid., p. 185.

10. Wallace Stegner, *Beyond the Hundredth Meridian: John Wesley Powell and the Second Opening of the West* (Lincoln: University of Nebraska Press, 1982), p. 236; subsequent pages are indicated in the text.
11. Paul Wallace Gates, *History of Public Land Law Development* (New York: Arno Press, 1979), p. 419.
12. U.S. Bureau of the Census, *Statistical Abstract of the United States: 1995* (Washington, D.C.: Government Printing Office, 1995), p. 17.

CHAPTER 8
Bioregionalist of the High and Dry: Stegner and Western Environmentalism

Previous versions of this essay have appeared in *Montana: The Magazine of Western History* 43 (Autumn 1993) and in Charles E. Rankin, ed., *Wallace Stegner: Man and Writer* (Albuquerque: University of New Mexico Press, 1996).

1. Wallace Stegner in *Under Western Skies: Writers and the West Calendar* (San Francisco: Browntrout Publishers, 1996), May excerpt.
2. The quotations are from T. H. Watkins, "Typewritten on Both Sides: The Conservation Career of Wallace Stegner," *Audubon* 89(5) (September 1987):88–103; reprinted as Chapter 10 in this volume.
3. My handling of the Dinosaur National Monument episode largely follows Mark Harvey's treatment in *A Symbol of Wilderness* (Albuquerque: University of New Mexico Press, 1994).
4. Wallace Stegner, "Wilderness Letter," in Frank Bergon, ed., *The Wilderness Reader* (New York: New American Library, 1980), p. 328.
5. Much of my interpretation here is based on a breakfast conversation with Stewart Udall in Missoula, Montana, in April 1996. Notes in possession of the author.
6. Wallace Stegner, *Where the Bluebird Sings to the Lemonade Springs: Living and Writing in the West* (New York: Penguin, 1992), p. 61.
7. I say this despite Stegner's contradictory claim, in the essay "Thoughts in a Dry Land" in *Where the Bluebird Sings* (p. 55), that "the West is less a place than a process." That sentence is preceded by this one, which clarifies his meaning: "The Westerner is less a person than a continuing adaptation." I think Stegner was speaking here about our continuing need to adapt socially to western aridity.
8. Stegner, *Where the Bluebird Sings*, pp. 46–47.
9. Larry Price, *Mountains and Man: A Study of Process and Environment* (Berkeley: University of California Press, 1981), p. 62; Thomas Vale, "Mountains and Moisture in the West," in William Wyckoff and Larry Dilsaver, eds., *The Mountainous West: Explorations in Historical Geography* (Lincoln: University of Nebraska Press, 1995), pp. 141–165.

10. These are the opening lines of Stegner's audiocassette, *A Sense of Place* (Louisville, Colo.: Audio Press, 1989).
11. Watkins, Chapter 10 in this volume.

CHAPTER 9
Wallace Stegner: Geobiographer

1. Wallace Stegner and Richard W. Etulain, *Conversations with Wallace Stegner on Western History and Literature* (Salt Lake City: University of Utah Press, 1983), p. 196.
2. Quoted in Leon Edel, *Writing Lives: Principia Biographica* (New York: Norton, 1987), p. 195.
3. Ibid., pp. 13–14.
4. Ibid., p. 17. Edel later restates the principle: "Every life takes its own form and a biographer must find the ideal and unique literary form that will express it" (p. 30).
5. Ibid., p. 145.
6. Ibid., p. 213.
7. Stegner, *One Way to Spell Man: Essays with a Western Bias* (Garden City, N.Y.: Doubleday, 1982), p. 94.
8. Stegner and Etulain, *Conversations*, p. 124.
9. Ibid., p. 125.
10. Ibid., p. 192.
11. Ibid., p. 87; emphasis added. For an excellent discussion of the problematic nature of *Joe Hill* see Robert H. Keller, "'Joe Hill Ain't Never Died': Wallace Stegner's Act of Literary Imagination," in Charles E. Rankin, ed., *Wallace Stegner: Man and Writer* (Albuquerque: University of New Mexico Press), pp. 163–179.
12. Stegner and Etulain, *Conversations*, p. 68.
13. Keller, "'Joe Hill Ain't Never Died,'" p. 164.
14. Stegner and Etulain, *Conversations*, p. 71.
15. Ibid., p. 72.
16. Wallace Stegner, *The Big Rock Candy Mountain* (New York: Penguin, 1991), p. 464.
17. Wallace Stegner, *Where the Bluebird Sings to the Lemonade Springs: Living and Writing in the West* (New York: Penquin, 1992) pp. 4, 19–20.
18. Edel, *Writing Lives*, p. 176.
19. Stegner and Etulain, *Conversations*, p. 162.
20. Edel, *Writing Lives*, p. 202.
21. Wallace Stegner, *Beyond the Hundredth Meridian* (New York: Penguin, 1992), p. 303.

22. Ibid., pp. 116–117.
23. Although, as Jackson Benson points out in *Wallace Stegner: His Life and Work*, (New York: Viking, 1996), De Voto's acerbic tone in the introduction may well have cost Stegner a Pulitzer Prize.
24. Edel, *Writing Lives*, p. 181.
25. Wallace Stegner, *The Uneasy Chair: A Biography of Bernard De Voto* (Salt Lake City: Peregrine Smith Books, 1988), p. 215.
26. Stegner and Etulain, *Conversations*, p. 90. Later in the interview Stegner says: "I was a good deal more aware of the fact that somehow these wanderings tie together elements of the national life, that the East and the West are both in it. The Midwest, unfortunately, isn't much in it, but the East and West and Mexico and lots of places are, and different classes of society, so that in many ways *Angle of Repose* is a book with a lot more range than *The Big Rock Candy Mountain*. I think it may be historically more significant for that reason" (p. 43).
27. Stegner, *The Uneasy Chair*, p. 3; subsequent pages are cited in the text.
28. Stegner, *Where the Bluebird Sings*, pp. 224 and 227.
29. Jackson Benson, "Finding a Voice of His Own: The Story of Wallace Stegner's Fiction," in Charles E. Rankin, ed., *Wallace Stegner: Man and Writer* (Albuquerque: University of New Mexico Press, 1996), p. 208. First published in *Western American Literature* 29 (Summer 1994):99–122.
30. Ibid., p. 215.
31. Ibid., p. 218.
32. Ibid., p. 227.
33. Stegner and Etulain, *Conversations*, p. 166.
34. Quoted in Edel, *Writing Lives*, p. 68.

CHAPTER 10
Reluctant Tiger: Wallace Stegner Takes Up the Conservation Mantle

1. See T. H. Watkins, "Typewritten on Both Sides: The Conservation Career of Wallace Stegner," *Audubon* 89(5) (September 1987):88–103; "Bearing Witness for the Land: The Conservation Career of Wallace Stegner," *South Dakota Review* 23 (Winter 1985):42–57.
2. Wallace Stegner, *The Big Rock Candy Mountain* (Lincoln: University of Nebraska Press, 1983), pp. 463–464.
3. Wallace Stegner, *The Spectator Bird* (Garden City, N.Y.: Doubleday, 1976), p. 213.
4. Wallace Stegner, *A Shooting Star* (New York: Viking Press, 1961), p. 321.

5. Wendell Berry, "Wallace Stegner and the Great Community," *South Dakota Review* 23 (Winter 1985):16; reprinted in Wendell Berry, *What Are People For?* (San Francisco: North Point, 1990), pp. 48–57.
6. Wallace Stegner, *Wolf Willow: A History, a Story, and a Memory of the Last Plains Frontier* (Lincoln: University of Nebraska Press, 1980), p. 7.
7. May Sarton, "Boulder Dam," in *May Sarton: Collected Poems (1930–1973)* (New York: Norton, 1974), p. 51.
8. Stegner, "Rediscovering America, Part II," *Saturday Review* August 24, 1946, p. 12.
9. Wallace Stegner, "Battle for the Wilderness," *The New Republic,* February 15, 1954, p. 15.
10. Wallace Stegner, *Beyond the Hundredth Meridian: John Wesley Powell and the Second Opening of the West* (New York: Penguin, 1992), p. 88.
11. Stephen Fox, *John Muir and His Legacy: The American Conservation Movement* (Boston: Little, Brown, 1981), p. 286.
12. Wallace Stegner and Richard W. Etulain, *Conversations with Wallace Stegner on Western History and Literature* (Salt Lake City: University of Utah Press, 1983), pp. 169–170.
13. Wallace Stegner, "The Geography of Hope: Saga of a Letter," *Living Wilderness* 44(151) (December 1980):12.
14. Wallace Stegner, *The Sound of Mountain Water: The Changing American West* (Garden City, N.Y.: Doubleday, 1969), p. 153.
15. Wallace Stegner, *Angle of Repose* (Garden City, N.Y.: Doubleday, 1971), 229–231.
16. Wallace Stegner, *The Uneasy Chair: A Biography of Bernard De Voto* (Salt Lake City: Peregrine Smith Books, 1988), p. 380.
17. Stegner, *Sound of Mountain Water*, p. 134.
18. Ibid., pp. 18–19.
19. Ibid., p. 38.
20. Wallace Stegner, "Living on Our Principal," *Wilderness* 48(168) (Spring 1985):21; reprinted in J. Baird Callicott, ed., *Companion to* A Sand County Almanac (Madison: University of Wisconsin Press, 1987), pp. 233–245.
21. Wallace Stegner and Page Stegner, *American Places* (New York: Dutton, 1981), p. 217.

Chapter 11
Nature and People in the American West: Guidance from Wallace Stegner's Sense of Place

1. Wallace Stegner, *The Big Rock Candy Mountain* (New York: Pocket Books, 1977), pp. 516, 524.

2. Wallace Stegner, *Where the Bluebird Sings to the Lemonade Springs: Living and Writing in the West* (New York: Random House, 1992), p. 17; subsequent pages are cited in the text.
3. Wallace Stegner and Richard W. Etulain, *Conversations with Wallace Stegner on Western History and Literature* (Salt Lake City: University of Utah Press, 1983), p. 193.
4. Stegner, *Where the Bluebird Sings*, pp. 17–18.
5. Wallace Stegner, *Wolf Willow* (New York: Viking, 1973), pp. 275–276.
6. Ibid., p. 282.
7. Stegner, *Where the Bluebird Sings*, p. 49.
8. Stegner, *Wolf Willow*, p. 8.
9. Wallace Stegner, "Wilderness Letter," in David Brower, ed., *Wilderness: America's Living Heritage* (San Francisco: Sierra Club, 1961), p. 97. Reprinted in Wallace Stegner's *The Sound of Mountain Water: The Changing American West* (New York: Dutton, 1980).
10. John Wesley Powell, *Report on the Lands of the Arid Region of the United States*, Wallace Stegner, ed. (Cambridge, Mass.: Belknap Press, 1962), p. xxiv; Stegner, *Where the Bluebird Sings*, p. 61.
11. Michael Solot, "Carl Sauer and Cultural Evolution," *Annals of the Association of American Geographers* 76 (1986):508–520; Ronald Abler, Melvin Marcus, and Judy Olson, eds., *Geography's Inner Worlds* (New Brunswick, N.J.: Rutgers University Press, 1992); Stephen Frenkel, "Geography, Empire, and Environmental Determinism," *Geographical Review* 82(3) (1992):143–153.
12. Donald Worster, *Dust Bowl: The Southern Plains in the 1930s* (New York: Oxford University Press, 1979), p. 197.
13. Stephen Frenkel, "Old Theories in New Places? Environmental Determinism and Bioregionalism," *Professional Geographer* 46(3) (1994):289–295.
14. Stegner, *Wolf Willow*, p. 282.
15. William Denevan, "The Pristine Myth: The Landscape of the Americas in 1492," *Annals of the Association of American Geographers* 82 (1992):369–385.
16. National Geographic Society, *Historical Atlas of the United States* (Washington, D.C.: National Geographic Society, 1993).
17. John Vankat, *The Natural Vegetation of North America* (New York: Wiley, 1979); Michael G. Barbour and William D. Billings, *North American Terrestrial Vegetation* (New York: Cambridge University Press, 1988).
18. Alfred L. Kroeber, *Cultural and Natural Areas of Native North America*, in University of California Publications in American Archaeology and Ethnology, vol. 28. (Berkeley: University of California Press, 1939); William Denevan, *The Native Population of the Americas in 1492*, 2nd ed. (Madison: University of Wisconsin Press, 1992).
19. Henry A. Wright and Arthur W. Bailey, *Fire Ecology* (New York: Wiley, 1982); U.S. Forest Service, "Proceedings—Symposium and Workshop on

Wilderness Fire." *General Technical Report INT-182* (Ogden, Utah: Intermountain Forest and Range Experiment Station, 1985).

20. Thomas Vale, "Wilderness Imagery in the New Ecology, the New Western History, and the New Myth of the Humanized Landscape; or, Why Some Old Ideas Are Really OK," public lecture, University of Wisconsin, 1995.
21. Stegner and Etulain, *Conversations*, p. 178.
22. Wallace Stegner, *Where the Bluebird Sings*, p. 128.
23. Michael Pollon, *Second Nature* (New York: Dell, 1991); William Cronon, "The Trouble with Wilderness; or, Getting Back to the Wrong Nature," in William Cronon, ed., *Uncommon Ground: Toward Reinventing Nature* (New York: Norton, 1995), pp. 69–90.
24. Thomas Vale, "Conservation Strategies in the Redwoods," *Yearbook of the Association of Pacific Coast Geographers* 36 (1974):103–112; Susan Schrepfer, "Perspectives on Conservation: Sierra Club Strategies in Mineral King," *Journal of Forest History* 20 (1976):176–190; Susan Schrepfer, "Conflict in Preservation: The Sierra Club, Save-the-Redwoods League, and Redwood National Park." *Journal of Forest History* 24 (1980):60–77.
25. Stegner and Etulain, *Conversations*, p. 197.
26. Wallace Stegner, "History, Myth, and the Western Writer," in J. Golden Taylor, ed., *Great Western Short Stories* (Palo Alto: American West, 1967), pp. xxiii–xxiv; reprinted in Stegner, *The Sound of Mountain Water*.
27. See, for example, Patricia Nelson Limerick, Clyde A. Milner II, and Charles E. Rankin, eds., *Trails: Toward a New Western History* (Lawrence: University Press of Kansas, 1991); Cronon, "The Trouble with Wilderness"; Worster, *Dust Bowl*.
28. Patricia Nelson Limerick, *The Legacy of Conquest: The Unbroken Past of the American West* (New York: Norton, 1987), p. 25.
29. Stegner, *Where the Bluebird Sings*, p. 203.
30. William Cronon, "A Place for Stories: Nature, History, and Narrative," *Journal of American History* 78 (1992):1347–1376.
31. William Cronon, "Kennecott Journey: The Paths Out of Town," in William Cronon, George Miles, and Jay Gitlin, eds., *Under an Open Sky: Rethinking America's Western Past*, (New York: Norton, 1992), pp. 28–51; Nancy Langston, *Forest Dreams, Forest Nightmares: The Paradox of Old Growth in the Inland West* (Seattle: University of Washington Press, 1995); Terry Tempest Williams, *Refuge: An Unnatural History of Family and Place* (New York: Pantheon, 1991); William Kittredge, *Hole in the Sky: A Memoir* (New York: Vintage, 1992).
32. Stegner, *Where the Bluebird Sings*, p. 206.
33. Yi-Fu Tuan, *Space and Place* (Minneapolis: University of Minnesota Press, 1977).
34. Wallace Stegner and Page Stegner, *American Places* (New York: Dutton, 1981), p. 118.

35. Wallace Stegner, *Recapitulation* (Garden City, N.Y.: Doubleday, 1979), p. 3.
36. See Thomas Vale and Geraldine Vale, *Western Images, Western Landscapes: Travels Along U.S. 89* (Tucson: University of Arizona Press, 1989).
37. Stegner, *Wolf Willow*, p. 306.
38. Wallace Stegner, *Angle of Repose* (Greenwich, Conn.: Fawcett Crest, 1971), p. 497.
39. Stegner and Etulain, *Conversations*, p. 185.
40. Stegner, *Big Rock Candy Mountain*, pp. 516, 524.

CHAPTER 12
Field Report from the New American West

1. Wallace Stegner, *Where the Bluebird Sings to the Lemonade Springs: Living and Writing in the West* (New York: Random House, 1992), pp. 205–206.
2. Ibid., p. 3.
3. Ibid., p. 4.
4. Wallace Stegner, *Collected Stories of Wallace Stegner* (New York: Penguin, 1991), p. 225.
5. Census Bureau figures between 1992 and 1993; from Richard Lamm, Jeff Gersh, and William Eldred, "The West at Risk," unpublished manuscript, University of Denver, 1994.
6. William H. Romme, "Pseudo-Rural Landscapes in the Mountain West" in Joan Nassauer and Deborah Karasov, eds. *Placing Nature: Culture and Landscape Ecology* (Washington, D.C.: Island Press, 1997).
7. Jordan Bonfante, "The Sky's the Limit," *Time,* September 6, 1993, p. 23.
8. Morris E. Garnsey, *America's New Frontier* (New York: Knopf, 1950), p. 20.
9. Ibid., p. viii.
10. Bonfante, "The Sky's the Limit," p. 20.
11. Florence Williams, "Future Shock Hits Livingston," *High Country News*, April 5, 1993, pp. 14–15.
12. Lang Smith, "The Land Rush Is On," *Greater Yellowstone Report* 10(1) (1993): 1, 4–5.
13. John Cromartie, "Recent Demographic and Economic Changes in the West," statement before the U.S. House Committee on Natural Resources, Hearing on "The Changing Needs of the West,"April 7, 1994, USDA Economic Research Service; David M. Theobald, "Landscape Morphology and Effects of Mountain Development in Colorado: A Multi-Scale Analysis" (Ph.D. dissertation, University of Colorado, Boulder, 1995).
14. Colorado Department of Agriculture, *What Lies Ahead for Colorado's Ag Lands?* (Denver: Colorado Department of Agriculture, 1996).

15. Greater Yellowstone Coalition, *Sustaining Greater Yellowstone: A Blueprint for the Future* (Bozeman, Mont.: Greater Yellowstone Coalition, 1994).
16. Wallace Stegner and Page Stegner, *American Places* (New York: Dutton, 1981), pp. 109–110.
17. Ray Ring and Alexei Rubenstein, "Resort Towns Battle Monsters," *High Country News,* September 5, 1994, p. 8.
18. Ibid.
19. Richard L. Knight, George N. Wallace, and William E. Riebsame, "Ranching the View: Subdivisions Versus Agriculture," *Conservation Biology* 9(2) (1995):459–461.
20. Stegner and Stegner, *American Places*, p. 211.
21. Ibid.
22. Ibid., p. 216.
23. Smith, "The Land Rush Is On."
24. Mark Haggerty, "Fiscal Impact of Different Land Uses on County Government and School Districts in Gallatin County, Montana" (Bozeman: Local Government Center, Montana State University, 1996).
25. Stegner and Stegner, *American Places*, p. 112.
26. Ibid., pp. 111, 115–116.
27. Paul Larmer and Stephen Stuebner, "'Wise Use' Plans Abhor Change," *High Country News*, September 5, 1994, pp. 16–17; Paul Larmer and Ray Ring, "Can Planning Rein in a Stampede?" *High Country News*, September 5, 1994, pp. 6–8.
28. Stegner and Stegner, *American Places*, p. 215.
29. Ibid.
30. Ibid., p. 165.
31. Ibid., p. 188.
32. Patricia Nelson Limerick, *The Legacy of Conquest: The Unbroken Past of the American West* (New York: Norton, 1988).
33. R. G. Walsh, J. R. McKean, and C. J. Mucklow, "Recreation Value of Ranch Open Space" (Fort Collins: Department of Agricultural and Resource Economics, Colorado State University, 1993).
34. Norman L. Christensen, Ann M. Bartuska, James H. Brown, Stephen Carpenter, Carla D'Antonio, Robert Francis, Jerry F. Franklin, James A. MacMahon, Reed F. Noss, David J. Parsons, Charles H. Peterson, Monica G. Turner, and Robert G. Woodmansee, "The Report of the Ecological Society of America Committee on the Scientific Basis for Ecosystem Management," *Ecological Applications* 6(3) (1996):665–691.
35. Wallace Stegner, *The American West as Living Space* (Ann Arbor: University of Michigan Press, 1987), pp. 42–43.
36. Stegner and Stegner, *American Places*, p. 217.

37. Ibid., p. 56.
38. Ibid., p. 60.
39. Stegner, *Where the Bluebird Sings*, pp. xxii–xxiii.

CHAPTER 13
Stegner and Contemporary Western Politics of the Land

1. Wallace Stegner, *The Sound of Mountain Water: The Changing American West* (New York: Dutton, 1980), p. 19.

CHAPTER 14
Wilderness Conversation

1. Aldo Leopold, *A Sand County Almanac and Sketches Here and There* (New York: Oxford University Press, 1949), pp. 188, 194, 199–200; Wallace Stegner, *The Sound of Mountain Water: The Changing American West* (New York: Dutton, 1980), pp. 152–153.
2. Stegner, *Sound of Mountain Water*, p. 153.
3. Wallace Stegner, Introduction to "Wilderness at the Edge," Utah Wilderness Coalition, n.d.
4. Wallace Stegner, unpublished journal entry, 1966; published here with the permission of Mary Page Stegner.
5. Stephen Trimble and Terry Tempest Williams, eds., *Testimony: Writers of the West Speak on Behalf of Utah Wilderness* (Minneapolis: Milkweed Editions, 1996), p. 18.
6. Ibid., p. 46.
7. Ibid., p. 36.
8. Ibid., p. 63.
9. Ibid., p. 71.

ABOUT THE CONTRIBUTORS

Jackson J. Benson is professor of English at San Diego State University, where he teaches modern American literature. Benson's biography *The True Adventures of John Steinbeck, Writer* received the PEN-West USA Award for nonfiction in 1985. His biography of Stegner, *Wallace Stegner: The Man and His Work*, was published in 1996.

Dorothy Bradley served for many years in the Montana State Legislature and is currently director of the Water Resources Center of the Montana University System in Bozeman. She has also led the effort to establish a Wallace Stegner chair at Montana State University.

John Daniel is an Oregon poet, essayist, and environmental journalist who teaches in workshops and writer-in-residence positions around the country. His books include *The Trail Home: Nature, Imagination, and the American West* (1994), *Looking After: A Son's Memoir* (1996), and two collections of poems, *Common Ground* (1988) and *All Things Touched by Wind* (1994). He and his wife lived on the Stegner property for five years in the 1980s as renters and friends.

Dan Flores holds the A. B. Hammond Chair of Western History at the University of Montana in Missoula. His books include *Journal of an Indian Trader: Anthony Glass and the Texas Trading Frontier, 1790–1810* (1985) and *Caprock Canyon Lands: Journeys into the Heart of the Southern Plains* (1990).

Melody Graulich is professor of English and women's studies at the University of New Hampshire in Durham. She has received National Endowment for the Humanities and Huntington Library fellowships for her work on the literature of the American West. Graulich has edited three books by Mary Austin and is currently editing a volume of essays on Austin.

James R. Hepworth is director of the Confluence Press and associate professor of literature and languages at Lewis-Clark State College in Lewiston, Idaho. He has coedited two volumes: *The Stories That Shape Us* (1995) with Teresa Jordan and *Resist Much, Obey Little: Some Notes on Edward Abbey* (1996) with Gregory McNamee.

PAUL W. JOHNSON is chief of the Natural Resources Conservation Service in Washington, D.C. He formerly served in the Iowa State Legislature, and his family maintains a dairy farm in northeastern Iowa.

RICHARD L. KNIGHT is professor of wildlife biology at Colorado State University. He has authored numerous scientific articles and recently edited three volumes for Island Press: *A New Century for Natural Resources Management* (dedicated to Wallace Stegner), *Wildlife and Recreationists,* and *Stewardship Across Boundaries.*

CURT MEINE is a writer and conservation biologist with the International Crane Foundation in Baraboo, Wisconsin. He is the author of *Aldo Leopold: His Life and Work* (1988) and co-editor (with George W. Archibald) of *The Cranes: Status Survey and Conservation Action Plan* (1996).

WALTER NUGENT is professor of history at the University of Notre Dame, where he teaches American western and environmental history. His most recent book, *Crossings: The Great Transatlantic Migrations 1870–1914,* appeared in paperback in 1995.

THOMAS R. VALE is professor of geography and environmental studies at the University of Wisconsin-Madison, where he teaches courses in physical geography, the history of conservation and wilderness protection, and vegetation change. His books include *Progress Against Growth: Daniel B. Luten on the American Landscape* (1986) and, in conjunction with Geraldine Vale, *Western Images, Western Landscapes: Travels Along US 89* (1990) and *Time and the Tuolomne Landscape: Continuity and Change in the Yosemite High Country* (1994).

T. H. WATKINS is vice president of The Wilderness Society and served for many years as editor of its magazine *Wilderness.* Watkins has written prolifically on the history and landscape of the American West. His books include the award-winning *Righteous Pilgrim: The Life and Times of Harold Ickes* (1992), *Great Depression: America in the 1930s* (1993), and *World of Wilderness: Essays on the Power and Purpose of Wild Country* (1995). He is currently working on a biography of Stegner.

ELLIOTT WEST is professor of history at the University of Arkansas in Fayetteville. Among his published works are *The Saloon on the Rocky Mountain Mining Frontier* (1979) and *Growing Up with the Country: Childhood on the Far Western Frontier* (1989). His most recent book is *The Way to the West: Essays on the Central Plains* (1995).

CHARLES WILKINSON is professor of law at the University of Colorado in Boulder. He has explored the connections among the law, environmental policy, western history, geography, and literature in such books as *American Indians, Time and the Law: Native Societies in a Modern Constitutional Democracy* (1987), *Crossing the Next Meridian: Land, Water, and the Future of the West* (1992), and *The Eagle Bird: Mapping a New West* (1992).

TERRY TEMPEST WILLIAMS is a lifelong resident of Salt Lake City and naturalist-in-residence at the Utah Museum of Natural History. Her award-winning books include *Pieces of White Shell: A Journey to Navajoland* (1984), *Coyote's Canyon* (1989), *Refuge: An Unnatural History of Family and Place* (1991), *An Unspoken Hunger* (1994), and *Desert Quartet* (1995). With Stephen Trimble she compiled the collection *Testimony: Writers of the West Speak on Behalf of Utah Wilderness* (1996).

INDEX

Island Press Board of Directors

DATE			